Undergraduate Texts in Mathematics

Undergraduate Texts in Mathematics

Apostol: Introduction to Analytic Number Theory.
1976. xii, 338 pages. 24 illus.

Armstrong: Basic Topology.
1983. xii, 260 pages. 132 illus.

Bak/Newman: Complex Analysis.
1982. x, 224 pages. 69 illus.

Banchoff/Wermer: Linear Algebra Through Geometry.
1983. x, 257 pages. 81 illus.

Childs: A Concrete Introduction to Higher Algebra.
1979. xiv, 338 pages. 8 illus.

Chung: Elementary Probability Theory with Stochastic Processes.
1975. xvi, 325 pages. 36 illus.

Croom: Basic Concepts of Algebraic Topology.
1978. x, 177 pages. 46 illus.

Fischer: Intermediate Real Analysis.
1983. xiv, 770 pages. 100 illus.

Fleming: Functions of Several Variables. Second edition.
1977. xi, 411 pages. 96 illus.

Foulds: Optimization Techniques: An Introduction.
1981. xii, 502 pages. 72 illus.

Franklin: Methods of Mathematical Economics. Linear and Nonlinear Programming. Fixed-Point Theorems.
1980. x, 297 pages. 38 illus.

Halmos: Finite-Dimensional Vector Spaces. Second edition.
1974. viii, 200 pages.

Halmos: Naive Set Theory.
1974, vii, 104 pages.

Iooss/Joseph: Elementary Stability and Bifurcation Theory.
1980. xv, 286 pages. 47 illus.

Jänich: Topology
1984. ix, 180 pages (approx.). 180 illus.

Kemeny/Snell: Finite Markov Chains.
1976. ix, 224 pages. 11 illus.

Lang: Undergraduate Analysis
1983. xiii, 545 pages. 52 illus.

Lax/Burstein/Lax: Calculus with Applications and Computing, Volume 1.
1976. xi, 513 pages. 170 illus.

LeCuyer: College Mathematics with A Programming Language.
1978. xii, 420 pages. 144 illus.

Macki/Strauss: Introduction to Optimal Control Theory.
1981. xiii, 168 pages. 68 illus.

Malitz: Introduction to Mathematical Logic: Set Theory - Computable Functions - Model Theory.
1979. xii, 198 pages. 2 illus.

Martin: The Foundations of Geometry and the Non-Euclidean Plane.
1975. xvi, 509 pages. 263 illus.

Martin: Transformation Geometry: An Introduction to Symmetry.
1982. xii, 237 pages. 209 illus.

Millman/Parker: Geometry: A Metric Approach with Models.
1981. viii, 355 pages. 259 illus.

continued after Index

Klaus Jänich

Topology

Translated by Silvio Levy

With 181 Illustrations

Springer-Verlag
New York Berlin Heidelberg Tokyo

Klaus Jänich
Fakultät für Mathematik
Universität Regensburg
8400 Regensburg
Federal Republic of Germany

Silvio Levy (*Translator*)
Mathematics Department
Princeton University
Princeton, NJ 08544
U.S.A.

AMS Subject Classification: 54-01

Library of Congress Cataloging in Publication Data
Jänich, Klaus.
Topology.
(Undergraduate texts in mathematics)
Translation of: Topologie.
Bibliography: p.
Includes index.
1. Topology. I. Title. II. Series.
QA611.J3513 1984 514 83-14495

Original German edition: *Topologie*, Springer-Verlag, Berlin, Heidelberg, New York, © 1980.

Typeset by Composition House Ltd., Salisbury, England.
Printed and bound by R. R. Donnelley & Sons, Harrisonburg, Virginia.
Printed in the United States of America.

9 8 7 6 5 4 3 2 1

ISBN 0-387-90892-7 Springer-Verlag New York Berlin Heidelberg Tokyo
ISBN 3-540-90892-7 Springer-Verlag Berlin Heidelberg New York Tokyo

Preface

This volume covers approximately the amount of point-set topology that a student who does not intend to specialize in the field should nevertheless know. This is not a whole lot, and in condensed form would occupy perhaps only a small booklet. Our aim, however, was not economy of words, but a lively presentation of the ideas involved, an appeal to intuition in both the immediate and the higher meanings.

I wish to thank all those who have helped me with useful remarks about the German edition or the original manuscript, in particular, J. Bingener, Guy Hirsch and B. Sagraloff. I thank Theodor Bröcker for donating his "Last Chapter on Set-Theory" to my book; and finally my thanks are due to Silvio Levy, the translator. Usually, a foreign author is not very competent to judge the merits of a translation of his work, but he may at least be allowed to say: I like it.

Regensburg, May 1983 KLAUS JÄNICH

Contents

Introduction

§1. What Is Point-Set Topology About?

It is sometimes said that a characteristic of modern science is its high—and ever increasing—level of specialization; every one of us has heard the phrase "only a handful of specialists ...". Now a general statement about so complex a phenomenon as "modern science" always has the chance of containing a certain amount of truth, but in the case of the above cliché about specialization the amount is fairly small. One might rather point to the great and ever increasing *interweaving* of formerly separated disciplines as a mark of modern science. What must be known today by, say, both a number theorist and a differential geometer, is much more, even relatively speaking, than it was fifty or a hundred years ago. This interweaving is a result of the fact that scientific development again and again brings to light hidden analogies whose further application represents such a great intellectual advance that the theory based on them very soon permeates all fields involved, connecting them together. Point-set topology is just such an analogy-based theory, comprising all that can be said in general about concepts related, though sometimes very loosely, to "closeness", "vicinity" and "convergence".

Theorems of one theory can be instruments in another. When, for instance, a differential geometer makes use of the fact that for each point and direction there is exactly one geodesic (which he does just about every day), he is

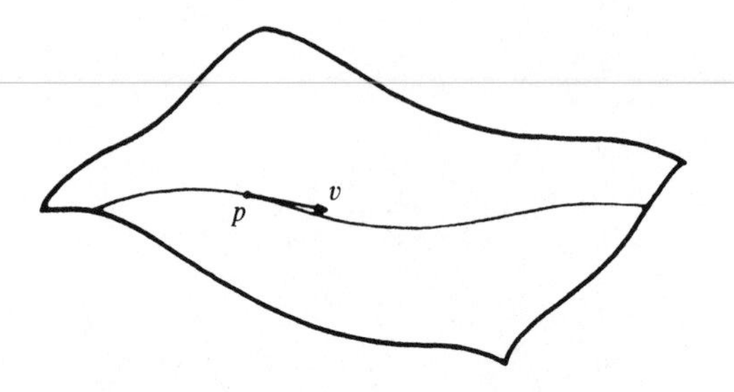

taking advantage of the Existence and Uniqueness Theorem for systems of second-order ordinary differential equations. On the other hand, the application of point-set topology to everyday uses in other fields is based not so much on deep theorems as on the unifying and simplifying power of its system of notions and of its felicitous terminology. And this power stems, in my understanding, from a very specific source, namely the fact that *point-set topology makes accessible to our spatial imagination a great number of problems which are entirely abstract and non-intuitive to begin with.* Many situations in point-set topology can be visualized in a perfectly adequate way in usual physical space, even when they do not actually take place there. Our spatial imagination, which is thus made available for mathematical reasoning about abstract things, is however a highly developed intellectual ability which is independent from abstraction and logical thinking; and this strengthening of our other mathematical talents is indeed the fundamental reason for the effectiveness and simplicity of topological methods.

§2. Origin and Beginnings

The emergence of fundamental mathematical concepts is almost always a long and intricate process. To be sure, one can point at a given moment and say: Here this concept, as understood today, is first defined in a clear-cut and plain way, from here on it "exists"—but by that time the concept had always passed through numerous preliminary stages, it was already known in important special cases, variants of it had been considered and discarded, etc., and it is often difficult, and sometimes impossible, to determine which mathematician supplied the decisive contribution and should be considered the originator of the concept in question.

In this sense one might say that the system of concepts of point-set topology "exists" since the appearance of Felix Hausdorff's book *Grundzüge der*

Mengenlehre (Leipzig, 1914). In its seventh chapter, "Point sets in general spaces", are defined the most important fundamental concepts of point-set topology. Maurice Fréchet, in his work "Sur quelques points du calcul fonctionnel" (*Rend. Circ. Mat. Palermo* **22**), had already come close to this mark, introducing the concept of metric spaces and attempting to grasp that of topological spaces as well (by axiomatizing the notion of convergence). Fréchet was primarily interested in function spaces and can perhaps be seen as the founder of the function analytic branch of point-set topology.

But the roots of the matter go, of course, deeper than that. Point-set topology, as so many other branches of mathematics, evolved out of the revolutionary changes undergone by the concept of geometry during the nineteenth century. In the beginning of the century the reigning view was the classical one, according to which geometry was the mathematical theory of the real physical space that surrounds us, and its axioms were seen as self-evident elementary facts. By the end of the century mathematicians had freed themselves from this narrow approach, and it had become clear that geometry was henceforth to have much wider aims, and should accordingly be made to work in abstract "spaces", such as n-dimensional manifolds, projective spaces, Riemann surfaces, function spaces etc. (Bolyai and Lobachevski, Riemann, Poincaré "and so on"—I'm not so bold as to try to delineate here this development process...). But now another contribution of paramount importance to the emergence of point-set topology was to be added to the rich variety of examples and the general ripeness to work with abstract spaces: namely, the work of Cantor. The dedication of Hausdorff's book reads: "To the creator of set theory, *Georg Cantor*, in grateful admiration."

"A topological space is a pair consisting of a set and a set of subsets, such that..."—it is indeed clear that the concept could never have been grasped in such generality were it not for the introduction of abstract sets in mathematics, a development which we owe to Cantor. But long before establishing his transfinite set theory Cantor had contributed to the genesis of point-set in an entirely diverse way, about which I would like to add something.

Cantor had shown in 1870 that two Fourier series that converge pointwise to the same limit function have the same coefficients. In 1871 he improved this theorem by proving that the coefficients have to be the same also when convergence and equality of the limits hold for all points outside a finite exception set $A \subset [0, 2\pi]$. In a work of 1872 he now dealt with the problem of determining for which *infinite* exception sets uniqueness would still hold.

An infinite subset of $[0, 2\pi]$ must of course have at least one cluster point:

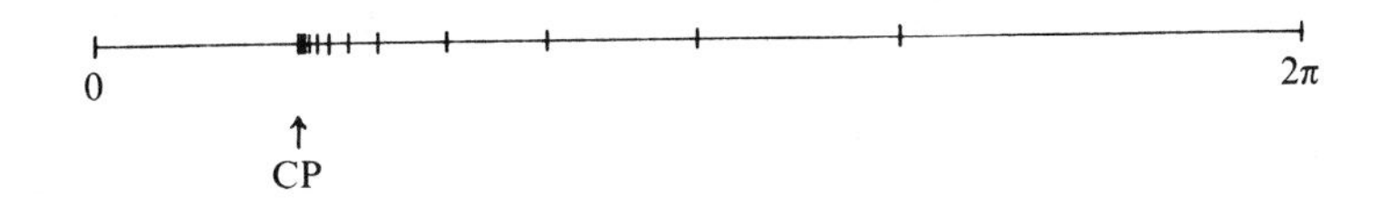

This is a very "innocent" example of an infinite subset of $[0, 2\pi]$. A somewhat "wilder" set would be one whose cluster points themselves cluster around some point:

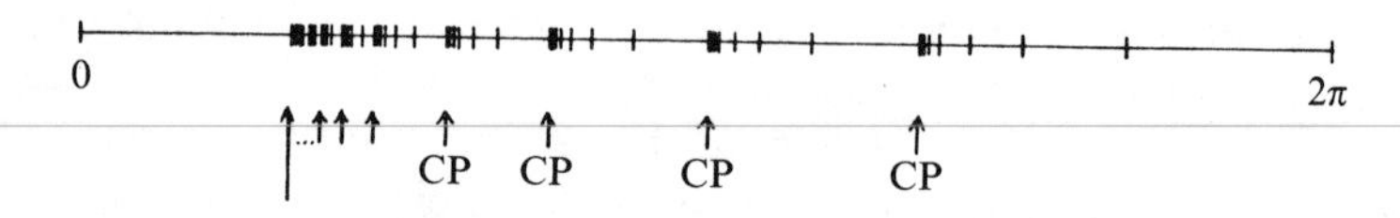

Cluster point of cluster points

Cantor now showed that if the sequence of subsets of $[0, 2\pi]$ defined inductively by $A^0 := A$ and $A^{n+1} := \{x \in [0, 2\pi] \mid x \text{ is a cluster point of } A^n\}$ breaks up after finitely many terms, that is if eventually we have $A^k = \varnothing$, then uniqueness *does* hold with A as the exception set. In particular a function that vanishes outside such a set (but not identically in the interval) cannot be represented by a Fourier series. This result helps to understand the strange convergence behavior of Fourier series, and the motivation for Cantor's investigation stems from classical analysis and ultimately from physics. But because of it Cantor was led to the discovery of a new type of subset $A \subset \mathbb{R}$, which must have been felt to be quite exotic, especially when the sequence $A, A^1, A^2, \ldots$ takes a *long* time to break off. Now the subsets of $\mathbb{R}$ move to the fore as objects to be studied in themselves, and, what is more, studied from what we would recognize today as being a topological viewpoint. Cantor continued along this path when later, while investigating general point sets in $\mathbb{R}$ and $\mathbb{R}^n$, he introduced the point-set topological approach, upon which Hausdorff could now base himself.

*

I do not want to give the impression that Cantor, Fréchet and Hausdorff were the only mathematicians to take part in the development and clarification of the fundamental ideas of point-set topology; but a more detailed treatment of the subject would be out of the scope of this book. I just wanted to outline, with a couple of sketchy but vivid lines, the starting point of the theory we are about to study.

CHAPTER I

Fundamental Concepts

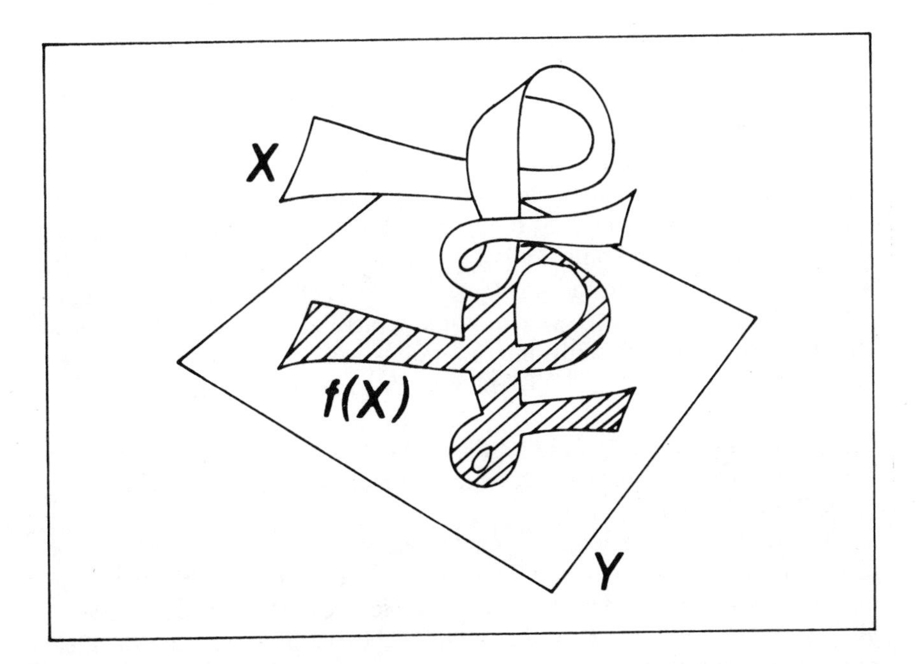

§1. The Concept of a Topological Space

Definition. A *topological space* is a pair $(X, \mathcal{O})$ consisting of a set X and a set $\mathcal{O}$ of subsets of X (called "open sets"), such that the following axioms hold:

Axiom 1. Any union of open sets is open.

Axiom 2. The intersection of any two open sets is open.

Axiom 3. $\varnothing$ and X are open.

One also says that $\mathcal{O}$ is the *topology* of the topological space $(X, \mathcal{O})$. In general one drops the topology from the notation and speaks simply of a topological space X, as we'll do from now on:

Definition. Let X be a topological space.

(1) $A \subset X$ is called *closed* when $X \backslash A$ is open.
(2) $U \subset X$ is called a *neighborhood* of $x \in X$ if there is an open set V with $x \in V \subset U$.

(3) Let $B \subset X$ be any subset. A point $x \in X$ is called an *interior, exterior* or *boundary* (or *frontier*) *point of* B, respectively, according to whether B, $X \backslash B$ or neither is a neighborhood of x.
(4) The set $\mathring{B}$ of the interior points of B is called the *interior* of B.
(5) The set $\bar{B}$ of the points of X which are not exterior points of B is called the *closure* of B.

These are then the basic concepts of point-set topology; and the reader who is being introduced to them for the first time should at this point work out a couple of exercises, in order to become familiar with them. Once, when I was still a student at Tübingen, I was grading some exercises after a lecture on these fundamental concepts. In the lecture it had already been established that a set is open if and only if all of its points are interior, and one exercise went like this: Show that the set of interior points of a set is always open. In came a student asking why we had not accepted his reasoning: "The set of interior points contains only interior points (an indisputable tautology); hence, the problem is trivial." There were a couple of other graders present and we all zealously tried to convince him that in talking about interior points you have to specify what set they are interior to, but in vain. When he realized what we wanted, he left, calmly remarking that we were splitting hairs. What could we answer?

Therefore, should among my readers be a complete newcomer to the field, I would suggest him to verify right now that the interior of B is the union of all open sets contained in B, and that the closure of B is the intersection of all closed sets containing B. And as food for thought during a peaceful afternoon let me add the following remarks.

Each of the three concepts defined above using open sets, namely, "closed sets", "neighborhoods" and "closure", can in its turn be used to characterize openness. In fact, a set $B \subset X$ is open if and only if $X \backslash B$ is closed, if and only if B is a neighborhood of each of its points, and if and only if $X \backslash B$ is equal to its closure. Thus the system of axioms defining a topological space must be expressible in terms of each one of these concepts, for instance:

Alternative Definition for Topological Spaces (Axioms for Closed Sets). A topological space is a pair $(X, \mathscr{A})$ consisting of a set X and a set $\mathscr{A}$ of subsets of X (called "closed sets"), such that the following axioms hold:

A1. Any intersection of closed sets is closed.
A2. The union of any two closed sets is closed.
A3. X and ϕ are closed.

This new definition is equivalent to the old in that $(X, \mathscr{O})$ is a topological space in the sense of the old definition if and only if $(X, \mathscr{A})$ is one in the sense of the new, where $\mathscr{A} = \{X \backslash V \mid V \in \mathscr{O}\}$. Had we given the second definition first, closedness would have become the primary concept, openness following by defining $X \setminus V$ to be open if and only if $V \subset X$ is closed. But the definition of concepts (2)–(5) would have been left untouched and given rise to the same system of concepts that we obtained in the beginning. It has become customary to start with open sets, but the idea of neighborhood is more intuitive, and indeed it was in terms of it that Hausdorff defined these notions originally:

Alternative Definition (Axioms for Neighborhood). A topological space is a pair $(X, \mathfrak{U})$ consisting of a set X and a family $\mathfrak{U} = \{\mathfrak{U}_x\}_{x \in X}$ of sets $\mathfrak{U}_x$ of subsets of X (called "neighborhoods of x") such that:

N1. Each neighborhood of x contains x, and X is a neighborhood of each of its points.
N2. If $V \subset X$ contains a neighborhood of x, then V itself is a neighborhood of x.
N3. The intersection of any two neighborhoods of x is a neighborhood of x.
N4. Each neighborhood of x contains a neighborhood of x that is also a neighborhood of each of its points.

One can see that these axioms are a bit more complicated to state than those for open sets. The characterization of topology by means of the closure operation, however, is again quite elegant and has its own name:

Alternative Definition (The Kuratowski Closure Axioms). A topological space is a pair $(X, \bar{\ })$ consisting of a set X and a map $\bar{\ }: \mathfrak{P}(X) \to \mathfrak{P}(X)$ from the set of all subsets of X into itself such that:

C1. $\overline{\varnothing} = \varnothing$.
C2. $A \subset \bar{A}$ for all $A \subset X$.
C3. $\bar{\bar{A}} = \bar{A}$ for all $A \subset X$.
C4. $\overline{A \cup B} = \bar{A} \cup \bar{B}$ for all $A, B \in X$.

Formulating what exactly the equivalence of all these definitions means and then proving it is, as we said, left as an exercise to the new reader. We will stick to our first definition.

§2. Metric Spaces

As we know, a subset of $\mathbb{R}^n$ is called open in the usual topology when every point in it is the center of some ball also contained in the set. This definition can be extended in a natural way if instead of $\mathbb{R}^n$ we consider a set X for which the notion of distance is defined; in particular every such space gives rise to a topological space. Let's recall the following

Definition (Metric Space). A metric space is a pair (X, d) consisting of a set X and a real function $d: X \times X \to \mathbb{R}$ (called the "metric"), such that:

M1. $d(x, y) \geq 0$ for all $x, y \in X$ and $d(x, y) = 0$ if and only if $x = y$.
M2. $d(x, y) = d(y, x)$ for all $x, y \in X$.
M3. (Triangle Inequality). $d(x, z) \leq d(x, y) + d(y, z)$ for all $x, y, z \in X$.

Definition (Topology of a Metric Space). Let (X, d) be a metric space. A subset $V \subset X$ is called open if for every $x \in V$ there is an $\varepsilon > 0$ such that the "ε-ball" $K_\varepsilon(x) := \{y \in X \mid d(x, y) < \varepsilon\}$ centered at x is still contained in V. The set $\mathcal{O}(d)$ of all open sets of X is called the topology of the metric space (X, d).

Then $(X, \mathcal{O}(d))$ is really a topological space: and here again our hypothetic novice has an opportunity to practice. But at this point even the more experienced reader could well lean back on his chair, stare at the void and think for a few seconds about what role is played here by the triangle inequality.

So? Well, absolutely none. But as soon as we want to start doing something with these topological spaces $(X, \mathcal{O}(d))$, the inequality will become very useful. It allows us, for example, to draw the conclusion, familiar from $\mathbb{R}^n$, that around each point y such that $d(x, y) < \varepsilon$ there is a small δ-ball entirely contained in the ε-ball around x:

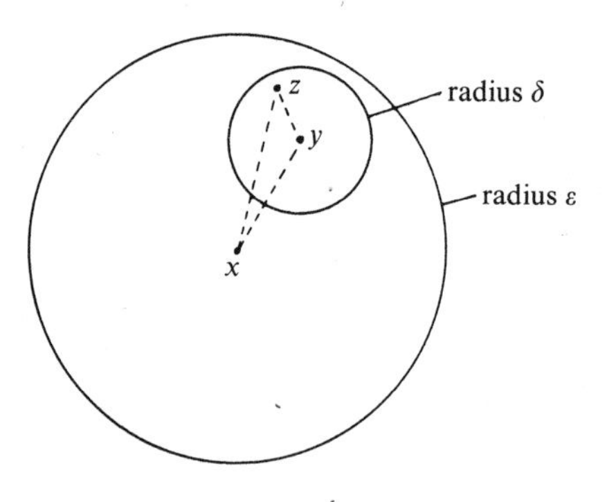

and consequently that the "open ball" $\{y \mid d(x, y) < \varepsilon\}$ is really open, whence in particular a subset $U \subset X$ is a neighborhood of x if and only if it contains a ball centered at x.

Metrics which are very different can in certain circumstances induce the same topology. If d and d' are metrics on X, and if every ball around x in the d metric contains a ball around x in the d' metric, we immediately have that every d-open set is d'-open, that is $\mathcal{O}(d) \subset \mathcal{O}(d')$. If furthermore the converse also holds, then the two topologies are the same: $\mathcal{O}(d) = \mathcal{O}(d')$. An example is the case $X = \mathbb{R}^2$ and

$$d(x, y) := \sqrt{(x_1 - y_1)^2 + (x_2 - y_2)^2}$$

$$d'(x, y) := \max\{|x_1 - y_1|, |x_2 - y_2|\}:$$

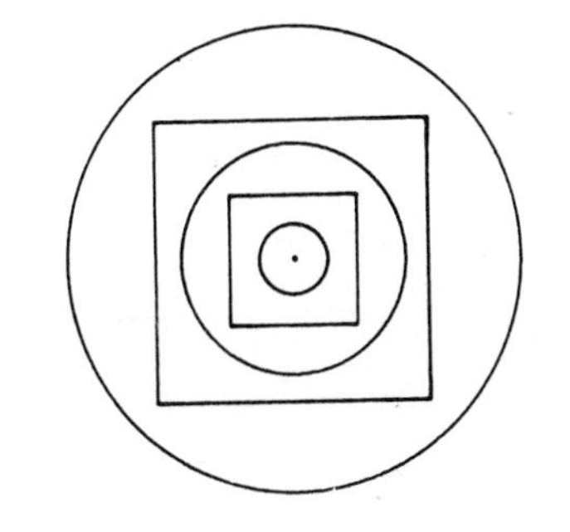

And here there is a simple but instructive trick that should be noted right from the start, a veritable talisman against false assumptions about the relationship between metric and topology: If (X, d) is a metric space, then so is (X, d'), where d' is given by $d'(x, y) := d(x, y)/(1 + d(x, y))$; moreover, as can be readily verified, $\mathcal{O}(d) = \mathcal{O}(d')$! But since all distances in d' are less than 1, it follows in particular that if a metric happens to be bounded this property can by no means be traced back to its topology.

Definition (Metrizable Spaces). A topological space $(X, \mathcal{O})$ is called *metrizable* if there is a metric d on X such that $\mathcal{O}(d) = \mathcal{O}$.

How can one determine whether or not a given topological space is metrizable? This question is answered by the "metrization theorems" of point-set topology. Are all but a few topological spaces metrizable, or is metrizability, on the contrary, a rare special case? The answer is neither, but rather the first than the second: there are a great many metrizable spaces. We will not deal with the metrization theorems in this book, but with the material in Chapters I, VI and VIII the reader will be quite well equipped for the further pursuit of this question.

§3. Subspaces, Disjoint Unions and Products

It often happens that new topological spaces are constructed out of old ones, and the three simplest and most important such constructions will be discussed now.

Definition (Subspace). If $(X, \mathcal{O})$ is a topological space and $X_0 \subset X$ a subset, the topology $\mathcal{O}|X_0 := \{U \cap X_0 | U \in \mathcal{O}\}$ on X_0 is called the *induced* or *subspace topology*, and the topological space $(X_0, \mathcal{O}|X_0)$ is called a *subspace* of $(X, \mathcal{O})$.

Instead of "open with respect to the topology of X_0" one says in short "open in X_0", and a subset $B \subset X_0$ is then open in X_0 if and only if it is the intersection of X_0 with a set open in X:

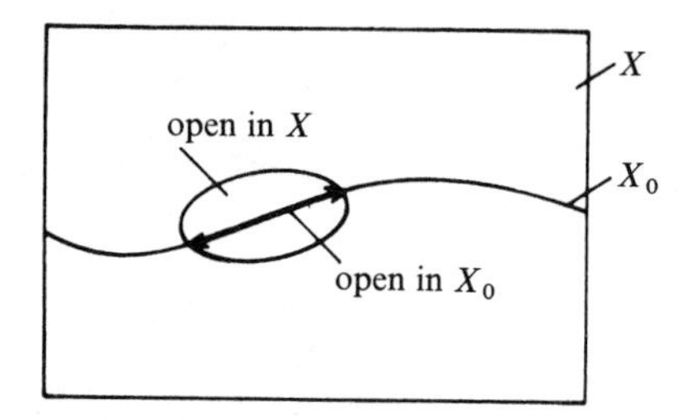

Thus such sets are not to be confused with sets "open and in X_0", since they *do not* have to be open—open, that is, in the topology of X.

Definition (Disjoint Union of Sets). If X and Y are sets, their *disjoint union* or *sum* is defined by means of some formal trick like for instance

$$X + Y := X \times \{0\} \cup Y \times \{1\}$$

—but we immediately start treating X and Y as subsets of $X + Y$, in the obvious way.

Intuitively this operation is nothing more than the disjoint juxtaposition of a copy of X and one of Y, and we obviously cannot write this as $X \cup Y$, since X and Y do not have to be disjoint to begin with, as for example when $X = Y$ and $X \cup X = X$ consists of only one copy of X.

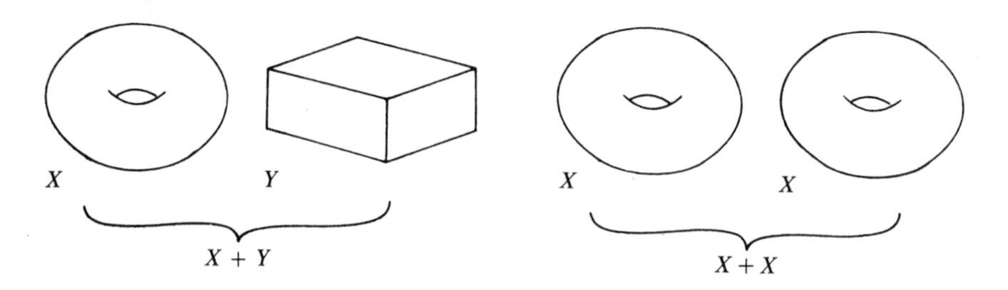

Definition (Disjoint Union of Topological Spaces). If $(X, \mathcal{O})$ and $(Y, \tilde{\mathcal{O}})$ are topological spaces, a new topology on $X + Y$ is given by

$$\{U + V | U \in \mathcal{O}, V \in \tilde{\mathcal{O}}\}$$

and the set $X + Y$ with this topology is called the *topological disjoint union* of the topological spaces X and Y.

Definition (Product Topology). Let X and Y be topological spaces. A subset $W \subset X \times Y$ is called *open in the product topology* if for each point $(x, y) \in W$ there are neighborhoods U of x in X and V of y in Y such that $U \times V \subset W$. The set $X \times Y$ endowed with the above topology is called the (Cartesian) product of the spaces X and Y.

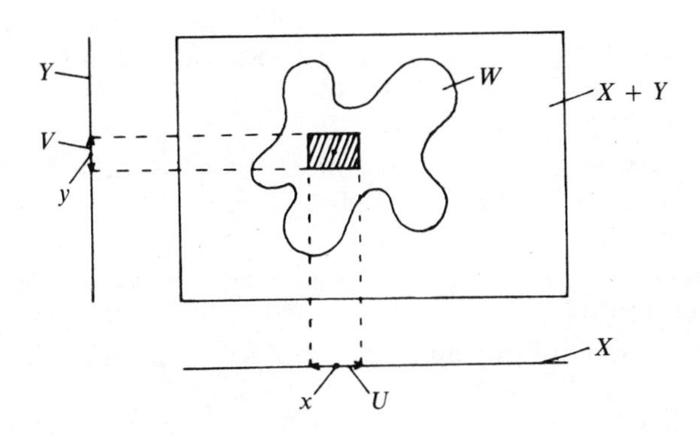

The box ☐ is the usual mental image for the Cartesian product of sets or topological spaces, and as long as we are dealing with nothing too complicated, this image is perfectly adequate. I will call the products

$$U \times V \subset X \times Y$$

of open sets $U \subset X$ and $V \subset Y$ *open boxes*. Open boxes are obviously open in the product topology, but they are not the only open sets: by themselves they do not form a topology, since the union of two boxes is not in general a box:

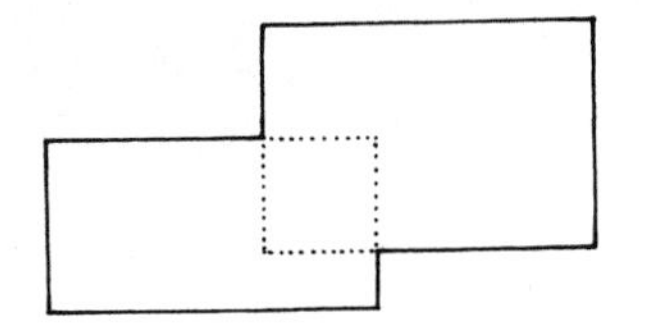

This trivial observation would not have occurred to me if I had not often come upon the opposite, erroneous, opinion, which must possess some peculiar attraction.—Well, that's it for the time being.

§4. Bases and Subbases

Definition (Basis). Let X be a topological space. A set $\mathfrak{B}$ of open sets is called a basis for the topology if every open set is a union of sets in $\mathfrak{B}$.

For example, the open boxes form a basis for the product topology, and the open balls in $\mathbb{R}^n$ form a basis for the usual topology in $\mathbb{R}^n$; but notice that the set of balls with rational radius and rational center coordinates (which is countable!) is also a basis for the topology of $\mathbb{R}^n$.

Definition (Subbasis). Let X be a topological space. A set $\mathfrak{S}$ of open sets is called a subbasis for the topology if every open set is a union of finite intersections of sets in $\mathfrak{S}$.

Of course the word "finite" here does not mean that the intersection should be a finite set, but that it is the intersection of finitely many sets. This includes the intersection of zero sets (that is, an empty family of sets), which by a meaningful convention is defined to be equal to the whole space (since in this way the formula $\bigcap_{\lambda\in\Lambda} S_\lambda \cap \bigcap_{\mu\in M} S_\mu = \bigcap_{\nu\in\Lambda\cup M} S_\nu$ still holds). Analogously, the union of an empty family of sets is suitably defined as the empty set.

With these conventions we then have that if X is a set and $\mathfrak{S}$ an arbitrary set of parts of X, there is exactly one topology $\mathcal{O}(\mathfrak{S})$ on X such that $\mathfrak{S}$ is a subbasis for $\mathcal{O}(\mathfrak{S})$ (the topology "generated" by $\mathfrak{S}$). It consists exactly of the unions of finite intersections of sets in $\mathfrak{S}$.

Thus a topology can be defined by prescribing a subbasis. But why should one want to do it? Well, it often happens that one wants a topology satisfying certain conditions. Usually one of these conditions refers to the *fineness* of the topology. If $\mathcal{O}$ and $\mathcal{O}'$ are topologies on X, and if $\mathcal{O} \subset \mathcal{O}'$ one says that $\mathcal{O}'$ is *finer* than $\mathcal{O}$ and that $\mathcal{O}$ is *coarser* than $\mathcal{O}'$; and often there are reasons to look for a topology which is as fine or as coarse as possible. To be sure, there is a coarsest topology on X, the so-called *trivial* topology, which contains only the sets X and $\varnothing$; and there is a finest topology, the so-called *discrete* topology, in which all subsets of X are open. But this is not enough, for one wishes to impose other conditions as well. In a typical case, the desired topology should on the one hand be as coarse as possible, and on the other contain at least the sets of $\mathfrak{S}$. There is always such a topology: it is exactly our $\mathcal{O}(\mathfrak{S})$.

§5. Continuous Maps

Definition (Continuous Map). Let X and Y be topological spaces. A map $f: X \to Y$ is called continuous if the inverse image of open sets is always open.

(1) There is at least *one* point with property P;
(2) If x has property P, the same applies to all points in a sufficiently small neighborhood;
(3) If x does not have property P, then the same applies to all points in a small neighborhood.

The following stronger concept is often of interest:

Definition (Path-Connectedness). X is said to be *path-connected* if every two points $a, b \in X$ are connected by a *path*, that is, a continuous map $\alpha: [0, 1] \to X$ such that $\alpha(0) = a$ and $\alpha(1) = b$:

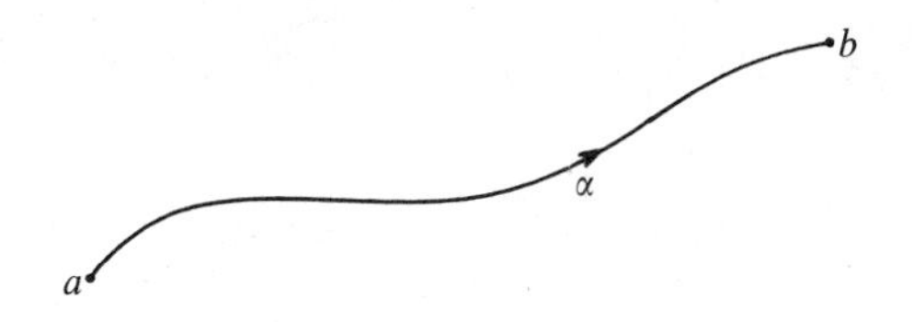

One sees immediately that a path-connected space X is connected: If $X = A \cup B$, with A and B open, non-empty and disjoint, there can be no path from $a \in A$ to $b \in B$, due to the connectedness of $[0, 1]$ (otherwise we would have $[0, 1] = \alpha^{-1}(A) \cup \alpha^{-1}(B)$ and so on).

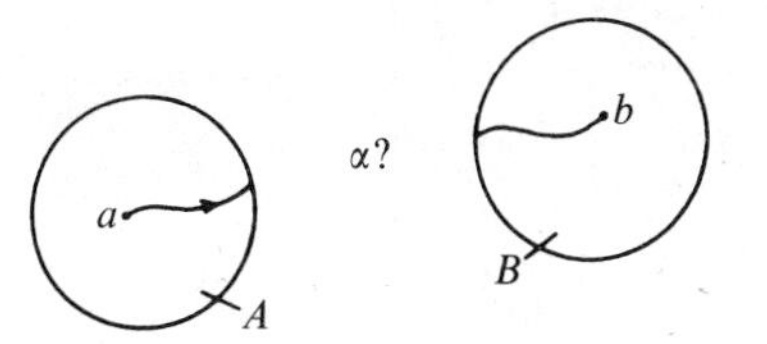

The converse is not true, though: a space can be connected and still manage to be "impassable" between two points. The subspace of $\mathbb{R}^2$ given by $\{(x, \sin \ln x) | x > 0\} \cup (0 \times [-1, 1])$ is an example:

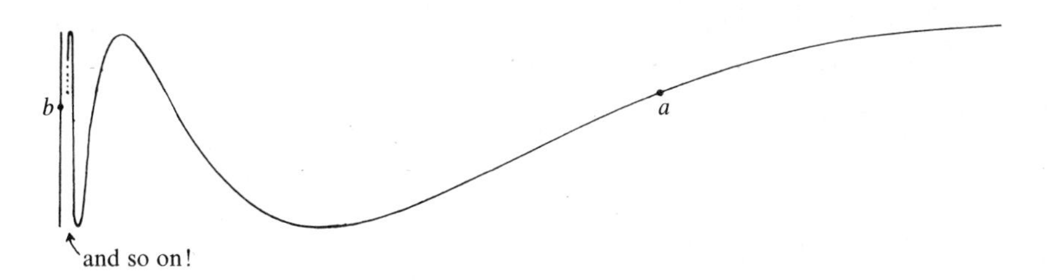

To conclude let me add three remarks concerning the behavior of connectedness under different operations. Topological properties such as connectedness tend to acquire, upon closer acquaintance, emotional overtones: some appear friendly and helpful, after we have seen several times how they make proofs easy or even possible in the first place; others, on the contrary, we come to dread, for the exactly opposite reason. True enough, a property of good repute can on occasion be an obstacle, and many properties are entirely ambivalent. But I can assure you that connectedness, Hausdorffness and compactness are predominantly "good" properties, and one would naturally like to know if such good properties are transferred from the building blocks to the final products by the usual topological constructions and processes. Thus:

Note 1. *Continuous images of (path-)connected spaces are (path-)connected. In other words, if X is (path-)connected and $f: X \to Y$ is continuous, then the subspace $f(X)$ of Y is also (path-)connected. For a decomposition of $f(X)$ as $A \cup B$ would imply the same for $X = f^{-1}(A) \cup f^{-1}(B)$, etc.*

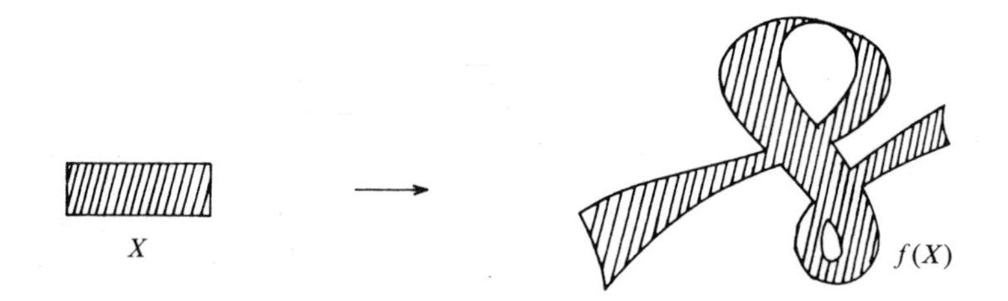

Note 2. *Non-disjoint unions of (path-)connected spaces are (path-)connected, that is if X_0 and X_1 are (path-)connected subspaces of X with $X = X_0 \cup X_1$ and $X_0 \cap X_1 \neq \varnothing$, then X is (path-)connected.*

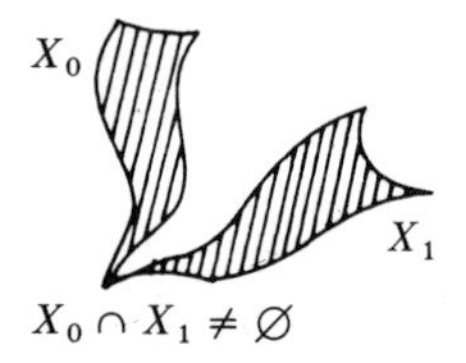

Note 3. *A Cartesian product $X \times Y$ of non-empty topological spaces X and Y is (path-)connected if and only if both factors are.*

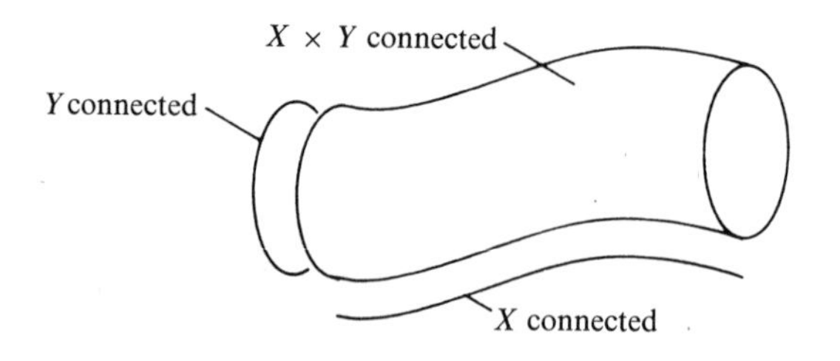

Facetious question: How about the disjoint union X and Y?

§7. The Hausdorff Separation Axiom

Definition (Hausdorff Separation Axiom). A topological space is called Hausdorff if for any two different points there exist disjoint neighborhoods.

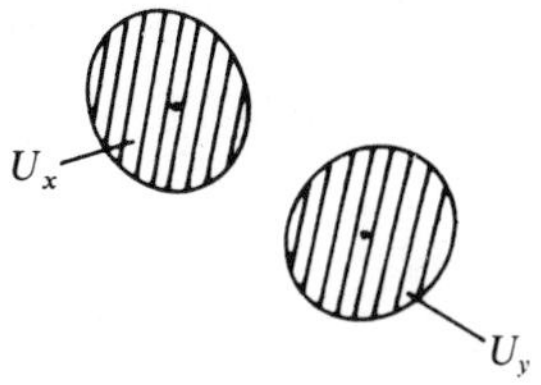

For example, every metric space is Hausdorff, for if d is a metric and $d(x, y) = \varepsilon > 0$, then the sets

$$U_x := \{z \mid d(x, z) < \varepsilon/2\} \quad \text{and} \quad U_y := \{z \mid d(y, z) < \varepsilon/2\},$$

for instance, are disjoint neighborhoods.

The property "non-Hausdorff" is quite counterintuitive and at first glance even unreasonable, seeming to go against our intuition of the neighborhood concept. For this reason Hausdorff included the above separation axiom in his original definition of "topological space" (1914). But later it was found that non-Hausdorff topologies too can be very useful, e.g. the "Zariski topology" in algebraic geometry. In any case one can step fairly deep into topology without really feeling a need for non-Hausdorff spaces, though here and there it is more convenient not to have to watch for Hausdorffness. For those who want to see such an exotic thing once, take a set X with more than one element and consider on it the trivial topology $\{X, \varnothing\}$.

One of the advantages offered by the separation axiom is the uniqueness of convergence:

Definition (Convergent Sequence). Let X be a topological space, $(x_n)_{n \in \mathbb{N}}$ a sequence in X. A point $a \in X$ is called limit of the sequence if for every neighborhood U of a there is an n_0 such that $x_n \in U$ for all $n \geq n_0$.

Note. *In a Hausdorff space a sequence can have at most one limit.*

In a trivial topological space, on the other hand, every sequence converges to every point.

As for behavior under operations, we note the following easily proved fact:

Note. *Every subspace of a Hausdorff space is Hausdorff, and two non-empty topological spaces X and Y are Hausdorff if and only if their disjoint union $X + Y$ is and if and only if their product $X \times Y$ is.*

*

The Hausdorff separation axiom is also called T_2. This sounds like there is a T_1, doesn't it? Well, how about this: $T_0, T_1, T_2, T_3, T_4, T_5$, not to mention $T_{2\frac{1}{2}}$ and $T_{3\frac{1}{2}}$! The Hausdorff axiom, however, is by far the most important of these, and deserves most to be kept in mind. Shall I say what T_1 stands for . . . ? But no. We can wait for it.

§8. Compactness

Ah, compactness! A wonderful property. This is true especially in differential and algebraic topology, as a rule, because everything works much more smoothly, easily and fully when we are dealing with compact spaces, manifolds, CW-complexes, groups etc. Now not everything in the world can be compact, but even for "non-compact" problems the compact case is often a good first step: We must first master the "compact terrain", which is easier to conquer, and then work our way into the non-compact case with modified techniques. Exceptions confirm the rule: Occasionally non-compactness also offers advantages, there is more "room" for certain constructions. . . . But now:

Definition (Compactness). A topological space is called compact if every open cover possesses a finite subcover. This means that X is compact if the following holds: If $\mathfrak{U} = \{U_\lambda\}_{\lambda \in \Lambda}$ is an arbitrary open cover of X, i.e. $U_\lambda \subset X$ open and $\bigcup_{\lambda \in \Lambda} U_\lambda = X$, then there are a finite number of $\lambda_1, \ldots, \lambda_r \in \Lambda$ such that $U_{\lambda_1} \cup \cdots \cup U_{\lambda_r} = X$.

(**Remark.** Many authors call such spaces "quasicompact" and save the word "compact" for "quasicompact and Hausdorff".)

In compact spaces the following type of generalization from "local" to "global" properties is possible: Let X be a compact space and P a property that the open subsets of X may or may not have, and also such that if U and V have it, then so does $U \cup V$. (Examples below.) Then if X has this property *locally*, i.e. every point has a neighborhood with property P, then X itself has property P. In fact, such open neighborhoods form an open cover $\{U_x\}_{x \in X}$ of X; but, choosing the x_i appropriately, we have

$$X = U_{x_1} \cup \cdots \cup U_{x_r},$$

and by assumption the property is inductively transferred to finite unions, qed.

Example 1. Let X be compact and $f: X \to \mathbb{R}$ locally bounded (continuous, for example). Then f is bounded.

Proposition 3. *Two non-empty spaces X and Y are both compact if and only if their disjoint union is, and if and only if their product is.*

Proof. (We'll prove only that the product of compact spaces is compact, which is the most interesting and relatively more difficult assertion. The converse follows from Proposition 1, and the statement about the disjoint union is trivial.) Let X and Y be compact and $\{W_\lambda\}_{\lambda\in\Lambda}$ an open cover of $X \times Y$.

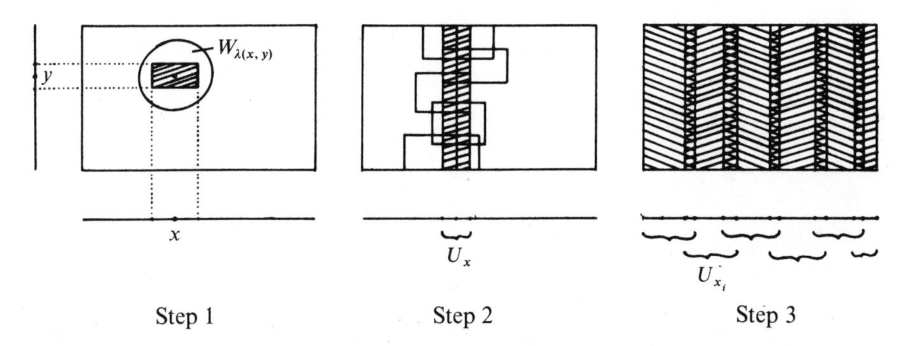

Step 1. We can choose for each (x, y) a $\lambda(x, y)$ such that $(x, y) \in W_{\lambda(x,y)}$, and because $W_{\lambda(x,y)}$ is open it contains an open box $U_{(x,y)} \times V_{(x,y)}$ around (x, y).

Step 2. For a fixed x the family $\{V_{(x,y)}\}_{y\in Y}$ is an open cover of Y, hence there are $y_1(x), \ldots, y_{r_x}(x)$ such that

$$V_{(x,y_1(x))} \cup \cdots \cup V_{(x,y_{r_x}(x))} = Y$$

Now put

$$U_{(x,y_1(x))} \cap \cdots \cap U_{(x,y_{r_x}(x))} =: U_x.$$

Step 3. Since X is compact, there are $x_1, \ldots, x_n$ with $U_{x_1} \cup \cdots \cup U_{x_n} = X$, and consequently $X \times Y$ is covered by the (finitely many!) $W_{\lambda(x_i, y_j(x_i))}$, $1 \leq i \leq n$, $1 \leq j \leq r_i$, qed. □

From the compactness of the closed interval and these three propositions we can prove the compactness of many other spaces, e.g. all closed subspaces of the n-dimensional cube and hence all closed and bounded subsets of $\mathbb{R}^n$. This is one half of the famous Heine–Borel theorem, which states that a subset of $\mathbb{R}^n$ is compact *if and only if* it is closed and bounded. Why is every compact subset X_0 of $\mathbb{R}^n$ closed and bounded? Well, we have already observed that continuous functions on compact sets are bounded, and this applies in particular to the norm function, hence X_0 is bounded. As for closedness, it follows from the following simple but useful

Lemma. *If X is a Hausdorff space and $X_0 \subset X$ a compact subspace, then X_0 is closed in X.*

Proof. We must show that $X \backslash X_0$ is open, hence that every point p has a neighborhood U that does not intersect X_0. For each $x \in X_0$ choose disjoint neighborhoods U_x of p and V_x of x. It may happen that U_x intersects X_0,

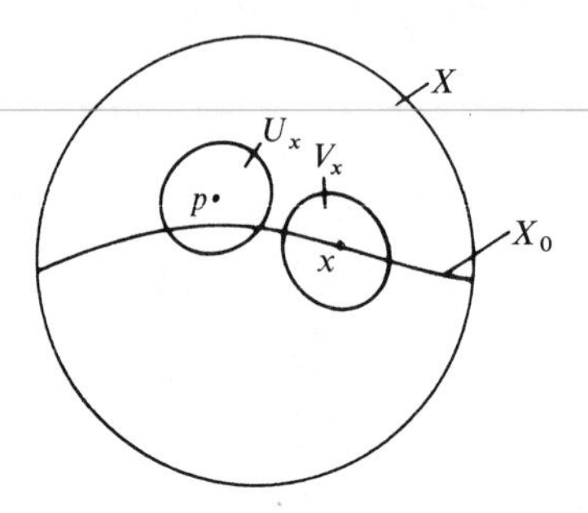

but at least it does not intersect the subset $V_x \cap X_0$, and if we now choose finitely many points $x_1, \ldots, x_n \in X_0$ such that

$$(V_{x_1} \cap X_0) \cup \cdots \cup (V_{x_n} \cap X_0) = X_0$$

(which is always possible because of compactness), then $U := U_{x_1} \cap \cdots \cap U_{x_n}$ is a neighborhood of p with the desired property of not intersecting X_0, qed. □

*

Last but not least, I will present a nice little theorem about homeomorphisms, but first a few words to put it in the proper light. The first notions of isomorphism are introduced to us in linear algebra, and to prove that a linear map $f: V \to W$ is an isomorphism, it is enough to verify bijectivity, because $f^{-1}: W \to V$ is then automatically linear. The same applies for instance to groups and group homomorphisms. Having got accustomed to that, it is with a certain chagrin that we realize that there are other nice properties of bijections which are not transferred to the inverse: for instance, $x \mapsto x^3$ defines a differentiable bijection from $\mathbb{R}$ into $\mathbb{R}$, but the inverse map is not differentiable at the origin:

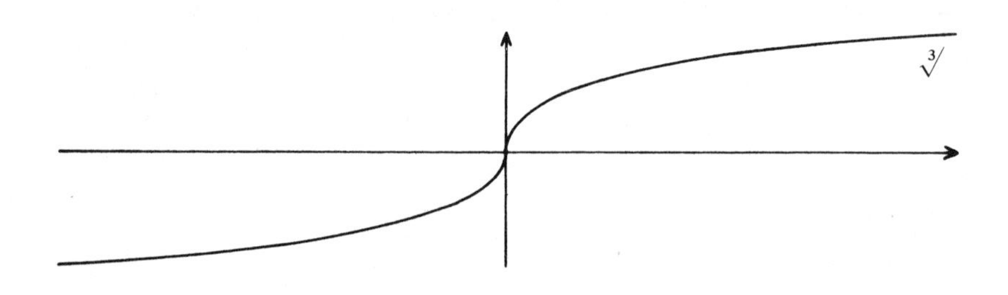

Unfortunately it is no better with continuity: take for instance the identity map from a set X with the discrete topology to X with the trivial topology. Nor does one have to resort to such extreme examples: Just wrap the half-open interval $[0, 2\pi)$ once around the unit circle, using the function $t \mapsto e^{it}$,

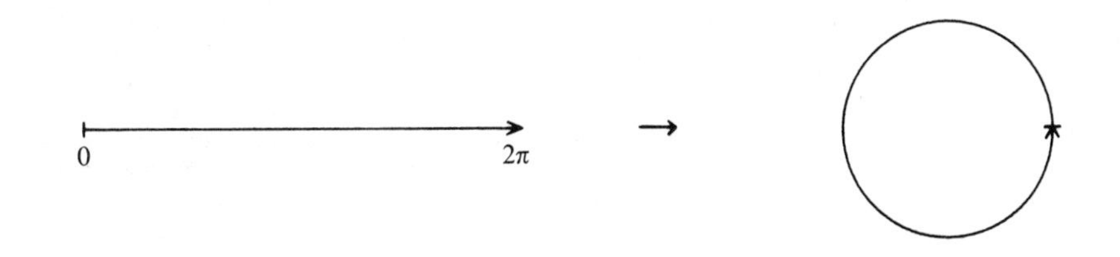

and we have a continuous bijection which cannot be a homeomorphism, because the circle is compact and the half-open interval is not. But even when f^{-1} *is* continuous, establishing this fact can turn out to be quite troublesome, especially when the continuity of f itself is obtained from an explicit formula $y = f(x)$, and there seems to be no way to write out a corresponding formula $x = f^{-1}(y)$. For this reason it is useful to have a condition, general in character and often satisfied, under which the inverse of a continuous bijection is always continuous:

Theorem. *A continuous bijection $f: X \to Y$ from a compact space X into a Hausdorff space Y is always a homeomorphism.*

PROOF. We have to show that the images of open sets are open, or, equivalently, that the images of closed sets are closed. Let then $A \subset X$ be closed. Then A is compact, since it is a closed subspace of a compact space; this means $f(A)$ is compact (continuous image of a compact space) and hence closed (compact subspace of the Hausdorff space Y), qed. □

CHAPTER II

Topological Vector Spaces

A large number of elements which intervene in mathematics are each completely determined by an infinite series of real or complex numbers: For example, a Taylor series is determined by the sequence of its coefficients . . . One can thus consider the numbers of the sequence which determine each of the elements as the coordinates of this element seen as a point of a space (E_ω) having a countably infinite number of dimensions. There are several advantages to working thus. First, the advantage that always appears when we use geometrical language, which favors intuition because of the analogies that it gives rise to . . .

MAURICE FRÉCHET
On Some Points of Functional Calculus (1906)

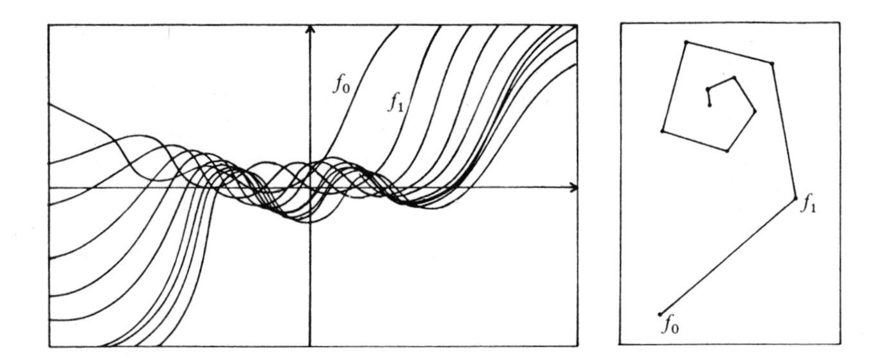

§1. The Notion of a Topological Vector Space

The present short chapter aims at nothing higher than presenting a certain class of examples of topological spaces, which really occur within the range of application of topology (in this case in functional analysis), and which is in fact of great significance: the topological vector spaces. It is only fair to place these examples right in the beginning, as they have played an important role in the formation of the notion of topological spaces (Fréchet 1906).

Definition (Topological Vector Space). Let $\mathbb{K} := \mathbb{R}$ or $\mathbb{C}$. A $\mathbb{K}$-vector space E with a topological space structure is called a *topological vector space* if its topological and linear structure are compatible in the following sense:

Axiom 1. The subtraction $E \times E \to E$ is continuous.
Axiom 2. Multiplication by scalars $\mathbb{K} \times E \to E$ is continuous.

Remark. Some authors impose an additional

Axiom 3. E is Hausdorff (e.g. Dunford–Schwartz [7]; but not Bourbaki [1]).

Instead of the subtraction we might as well have required the addition to be continuous, because it follows from Axiom 2 that the map $E \to E, x \mapsto -x$ is continuous, hence so is $E \times E \to E \times E$, $(x, y) \mapsto (x, -y)$. But there is one reason to phrase Axiom 1 with "subtraction" instead of "addition", and this reason, which I'll presently explain, is none the worse for being purely esthetic.

In the same way that there is a connection here between the notions of "vector space" and "topological space", so also many other interesting and useful concepts arise from a connection between the topology and the algebraic structure. In particular a group G which is also a topological space will be called a *topological group* if the group structure and the topology are "compatible". And what will be meant by that? Well, that the composition

$$G \times G \to G, \qquad (a, b) \mapsto ab$$

and the inverse map $G \to G$, $a \to a^{-1}$ are continuous. But these two conditions can be merged into one, the axiom for topological groups: The map $G \times G \to G$, $(a, b) \mapsto ab^{-1}$ is continuous.

Thus Axiom 1 says exactly that the additive group $(E, +)$ together with the topology of E forms a topological group.

In the next four paragraphs we will introduce the most common classes of topological vector space, in order of increasing generality.

§2. Finite-Dimensional Vector Spaces

$\mathbb{K}^n$, with the usual topology, is a topological vector space, and every isomorphism $\mathbb{K}^n \to \mathbb{K}^n$ is also a homeomorphism. Thus every n-dimensional vector space V has exactly one topology for which some (and consequently any) isomorphism $V \cong \mathbb{K}^n$ is a homeomorphism, and with this topology V becomes a topological vector space. This is all trivial, and undoubtedly the "usual" topology defined in this way is the most obvious one could find for V. But this topology is in fact more than just "obvious", for we have the following

Theorem (no proof given here, see, for instance, Bourbaki [1], Th. 2, p. 18). *The usual topology on a finite-dimensional vector space V is the only one that makes it into a Hausdorff topological vector space.*

The theorem shows that finite-dimensional topological vector spaces as such are not interesting, and the notion has been introduced because of the infinite-dimensional case. But even for these the theorem has an important consequence: namely, if V is a finite-dimensional vector subspace of any Hausdorff topological vector space E, then the topology on V induced from E is exactly the usual topology—even if E is one of the wilder specimens of its category.

§3. Hilbert Spaces

Let's recall that an *inner product space* is a real (resp. complex) vector space E together with a symmetric (resp. Hermitian) positive definite bilinear form $\langle .., .. \rangle$. Then for $v \in E$ the scalar $\|v\| := \sqrt{\langle v, v \rangle}$ is called the norm of v.

Note. *If $(E, \langle .., .. \rangle)$ is an inner product space, $d(v, w) := \|v - w\|$ defines a metric whose topology makes E into a topological vector space.*

Definition (Hilbert space). An inner product space is called a *Hilbert space* when it is complete relative to its metric, i.e. when every Cauchy sequence converges.

Hilbert spaces are surely, after finite-dimensional spaces, the most innocent topological vector spaces, and they can be completely classified, as follows: A family $\{e_\lambda\}_{\lambda \in \Lambda}$ of pairwise orthogonal unit vectors in a Hilbert space is called a *Hilbert basis* for H if the only vector orthogonal to all the e_λ is the zero vector. It can be proved that every Hilbert space has such a basis, any two bases of the same Hilbert space have the same cardinality, and finally two Hilbert spaces with equipotent bases are isometrically isomorphic.

§4. Banach Spaces

Definition (Normed Spaces). Let E be a $\mathbb{K}$-vector space. A map $\| .. \| : E \to \mathbb{R}$ is called a *norm* if the following three axioms hold:

N1. $\|x\| \geq 0$ for all $x \in E$, and $\|x\| = 0$ if and only if $x = 0$.
N2. $\|ax\| = |a| \|x\|$ for all $a \in \mathbb{K}$, $x \in E$.
N3. (Triangle Inequality). $\|x + y\| \leq \|x\| + \|y\|$ for all $x, y \in E$.

A pair $(E, \| .. \|)$ consisting of a vector space and a norm on it is called a *normed space.*

Note. *If $(E, \| .. \|)$ is a normed space, $d(x, y) := \|x - y\|$ defines a metric whose topology makes E into a topological vector space.*

Definition (Banach Space). A normed vector space is called a *Banach space* if it is complete, i.e. if every Cauchy sequence converges.

Hilbert and Banach spaces are, *in particular*, examples of topological vector spaces, but they have more structure than that: The scalar product $\langle .., .. \rangle$ or the norm $\| .. \|$ obviously cannot be recovered from the topology. Already for finite $n \geq 2$, a vector space V of dimension n can be endowed with many different norms which—in contrast with scalar products—cannot be obtained from one another by linear isomorphisms of the space into itself. Of course, all these norms define the same (i.e. the "usual") topology on V. Now, in *infinite* dimensions, even if one is only interested in the topological vector space structure (as often happens in functional analysis), Banach spaces define a very rich class of which it is difficult and perhaps impossible to get a complete overview.

§5. Fréchet Spaces

Definition (Seminorm). Let E be a $\mathbb{K}$-vector space. A map $|..|: E \to \mathbb{R}$ is called a seminorm if the following hold:

SN1. $|x| \geq 0$ for all $x \in E$.
N2. $|ax| = |a||x|$,
N3. Triangle inequality, } as for norms.

For example, $|..|_i: \mathbb{R}^n \to \mathbb{R}$, $x \mapsto |x_i|$ is a seminorm on $\mathbb{R}^n$.

We can talk about "open balls" for seminorms as well as for norms, and we will denote them by $B_\varepsilon(x) := \{y \in E \,|\, |x - y| < \varepsilon\}$; but in general there isn't anything "round" about them anymore.

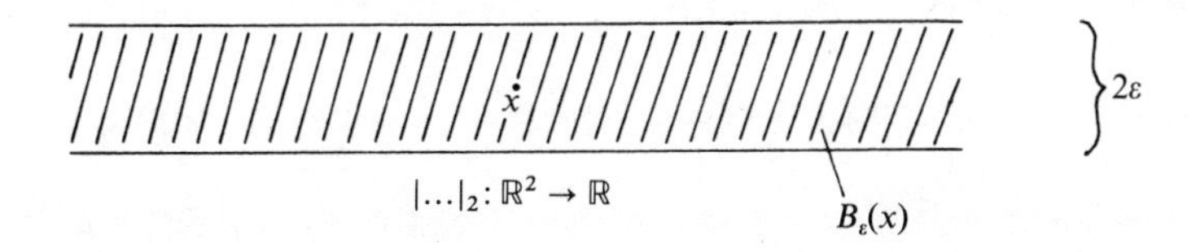

Definition. Let E be a vector space and $\{|..|_\lambda\}_{\lambda \in \Lambda}$ a family of seminorms on E. A subset $U \subset E$ is called open in the topology generated by the family of seminorms if every point of U belongs to a finite intersection of seminorm open balls which is contained in U; in other words, for every $x \in U$ there are $\lambda_1, \ldots, \lambda_r \in \Lambda$ and an $\varepsilon > 0$ such that $B_\varepsilon^{(\lambda_1)}(x) \cap \cdots \cap B_\varepsilon^{(\lambda_r)}(x) \subset U$.

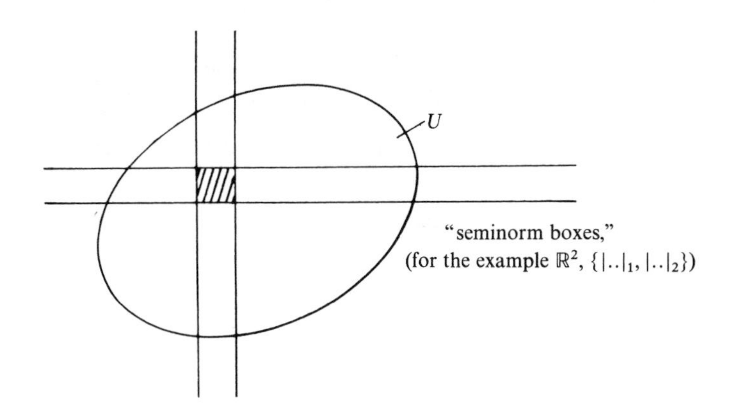

In the terminology of I, §4 these open balls of the seminorms $|..|_\lambda$, $\lambda \in \Lambda$ form a subbasis of, or generate, the topology.

Note. *With the topology given by the family of seminorms* $\{|..|_\lambda\}_{\lambda \in \Lambda}$, *E becomes a topological vector space, which moreover is Hausdorff if and only if* 0 *is the only vector for which all seminorms* $|..|_\lambda$ *are zero.*

Definition (Pre-Fréchet Space). A Hausdorff topological vector space whose topology can be defined by an at most countable family of seminorms is called a *pre-Fréchet space.*

Fréchet spaces will be the "complete" pre-Fréchet spaces. To be sure, completeness is a metric motion, but there is an obvious topological version of it for topological vector spaces:

Definition (Complete Topological Vector Spaces). A sequence $(x_n)_{n \geq 1}$ in a topological vector space is called a *Cauchy sequence* if for every neighborhood U of 0 there is an n_0 such that $x_n - x_m \in U$ for all $n, m \geq n_0$. If every Cauchy sequence converges, the space is called (*sequentially*) *complete.*

In normed spaces this concept of completeness is of course equivalent to the old one, obtained from the metric given by the norm.

Definition (Fréchet Space). A Fréchet space is a complete pre-Fréchet space.

Notice that pre-Fréchet spaces are always metrizable: If the topology is given by a sequence of seminorms $|..|_{n, n \geq 1}$, then

$$d(x, y) := \sum_{n=1}^{\infty} \frac{1}{2^n} \frac{|x - y|_n}{1 + |x - y|_n}$$

defines a metric which generates the same topology and for which the Cauchy sequences are the same.

§6. Locally Convex Topological Vector Spaces

Finally, let us define locally convex spaces, which are the most general class of topological vector spaces for which there exists a theory with decent theorems.

Definition. A topological vector space is called locally convex if every neighborhood of 0 contains a convex neighborhood of 0.

We mention the following facts to illustrate to what extent these spaces are more general than the preceding ones (no proof here; cf. [13] §18): A topological vector space is locally convex if and only if its topology can be given by a family of seminorms; and a locally convex topological vector space is a pre-Fréchet space if and only if it is metrizable.

§7. A Couple of Examples

Example 1. We consider the Lebesgue-integrable real functions f on $[-\pi, \pi]$ which satisfy

$$\int_{-\pi}^{\pi} f(x)^2 \, dx < \infty.$$

Two such functions will be called equivalent if they coincide outside a set of measure zero. The equivalence classes are called, somewhat loosely, square-integrable functions. Let H be the set of such functions. H has a canonical real vector space structure and can be made into a Hilbert space using, for instance, the following inner product:

$$\langle f, g \rangle := \frac{1}{\pi} \int_{-\pi}^{\pi} f(x)g(x) \, dx.$$

The trigonometric functions $e_k := \cos kx$, $e_{-k} := \sin kx$, $k \geq 1$, form, together with $e_0 := \sqrt{2}/2$, a Hilbert basis $\{e_n\}_{n \in \mathbb{Z}}$ for H, and the representation of elements $f \in H$ as $f = \sum_{n \in \mathbb{Z}} \langle f, e_n \rangle e_n$ is exactly the Fourier series of f.

Example 2. Let X be a topological space, $C(X)$ the vector space of bounded continuous functions on X, and $\|f\| := \sup_{x \in X} |f(x)|$. Then $(C(X), \|..\|)$ is a Banach space.

Example 3. Let $X \subset \mathbb{C}$ be a domain and $\mathcal{O}(X)$ the vector space of holomorphic functions on X, endowed with the topology given by the family

$$\{|..|_K\}_{K \subset X \text{ is compact}}$$

of seminorms $|f|_K := \sup_{z \in K} |f(z)|$ (topology of "compact convergence"). Then $\mathcal{O}(X)$ is a Fréchet space (we just have to consider a countable collection of K_n which "exhaust" X; completeness follows from the Weierstrass convergence theorem ...).

These are three out of a great number of "function spaces" which effectively come up in analysis. As mere vector spaces they did not have to be invented, they just are there and one can't miss them. And that the linear differential and integral operators behave as linear maps $L: E_1 \to E_2$ between function spaces also follows immediately from the nature of things. But mere linear algebra will lead us only to trivialities here; to understand the properties of these operators, we must study their continuity behavior under different topologies, and exploit our knowledge about the structure of abstract topological vector spaces. And while point-set topology, in whose praise I'm saying all this, does not exactly represent the cutting edge of research in the area of partial differential equations, it is nevertheless an indispensable instrument in it, to the point of being taken for granted.

I haven't yet given any examples of locally convex but non-metrizable, and hence non-pre-Fréchet, topological vector spaces. Well, such spaces also come up in a completely natural way in function analysis. For instance, it is sometimes necessary to consider the "weak topology" on a given topological vector space, that is, the coarsest topology for which all the old continuous linear maps $E \to \mathbb{R}$ (the "linear functionals") remain continuous, or in other words, the topology generated by $\{f^{-1}(U) | U \subset \mathbb{R}$ is open, $f: E \to \mathbb{R}$ is linear and continuous$\}$. With this topology E is still a topological vector space, but much more complicated than before. Even if we start with something as simple as an infinite-dimensional Hilbert space, we end up with a locally convex, Hausdorff, but non-metrizable topological vector space (cf. [4], p. 76).

CHAPTER III

The Quotient Topology

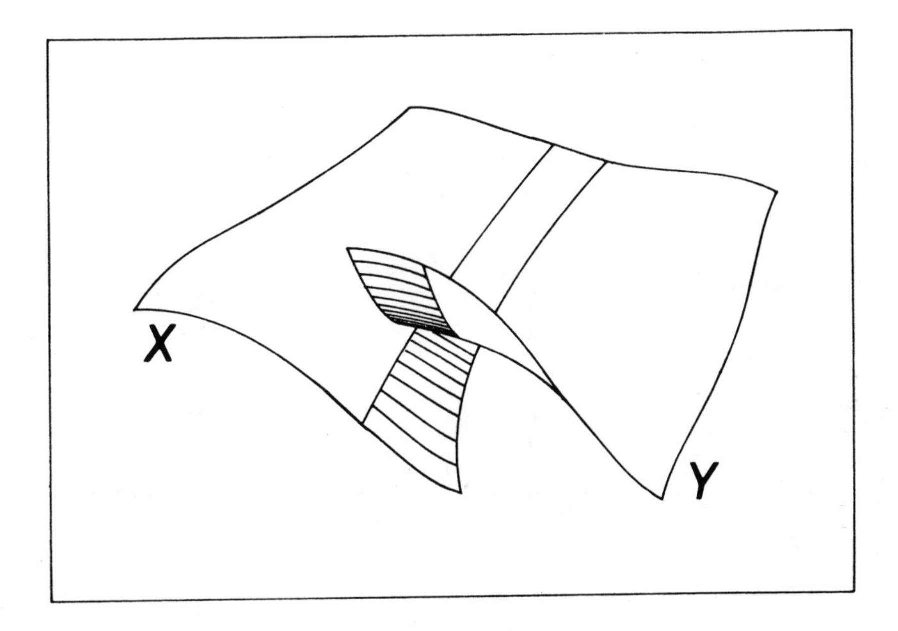

§1. The Notion of a Quotient Space

Notation. If X is a set and $\sim$ an equivalence relation on X, then $X/\sim$ will denote the set of equivalence classes, $[x] \in X/\sim$ the equivalence class of $x \in X$, and $\pi: X \to X/\sim$ the canonical projection, so that $\pi(x) := [x]$.

Definition (Quotient Space). Let X be a topological space and $\sim$ an equivalence relation on X. A set $U \subset X/\sim$ is called *open in the quotient topology* if $\pi^{-1}(U)$ is open in X. $X/\sim$, endowed with the topology thus defined, is called the *quotient of X by $\sim$*.

Note. *The quotient topology is obviously the finest topology on $X/\sim$ such that π is a continuous map.*

Just as we have, for the notions of subspace, disjoint union and product, a simple mental image on which we can base our intuition in the beginning, I would like to suggest a mental image for quotient spaces as well. In order to depict an equivalence relation, the best thing is to imagine the equivalence

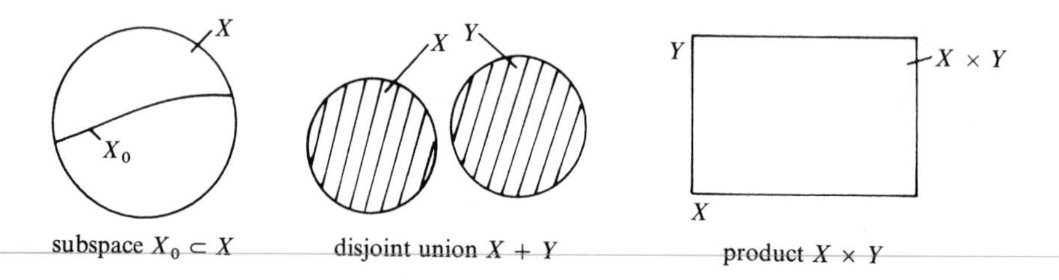

classes; but though these equivalence classes are the points of the quotient space, this is not enough, because our intuition demands a geometrical picture for the quotient space in which the points of the space are really "points" in the geometrical sense:

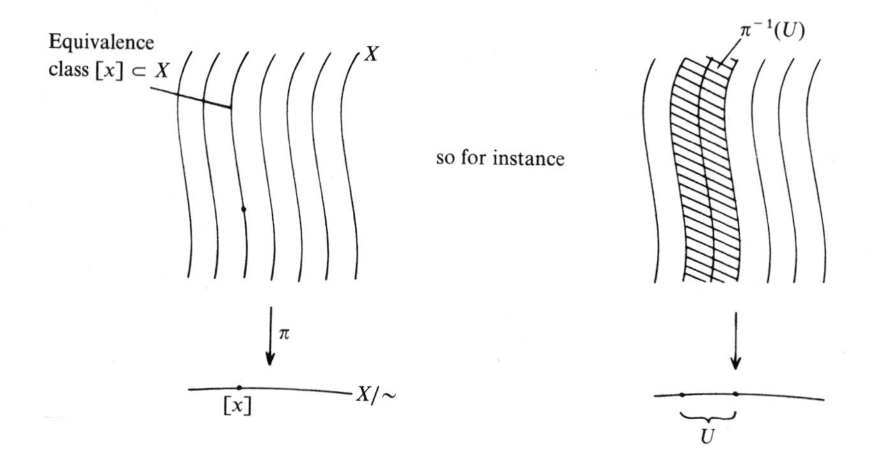

The next two sections contain all the "theory" of quotient spaces we need to know, and after that we are free to go into the really interesting part, that is, *examples* that really come up in mathematics, and not far-fetched contraptions.

§2. Quotients and Maps

Note 1 (Maps from Quotient Spaces). *Let Y be another topological space. A map $f : X/\sim \; \to Y$ is evidently continuous if and only if $f \circ \pi$ is continuous:*

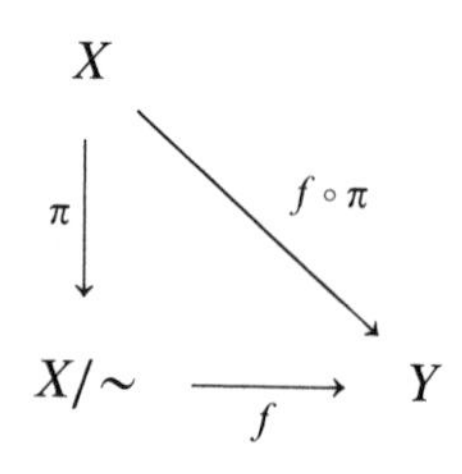

Note 2 (Maps into Quotient Spaces). There is no corresponding universal criterion for the continuity of a map $\varphi: Y \to X/\sim$, but the following trivial observation is often useful: *If there is a continuous map Φ: $Y \to X$ with*

$$\varphi = \pi \circ \Phi,$$

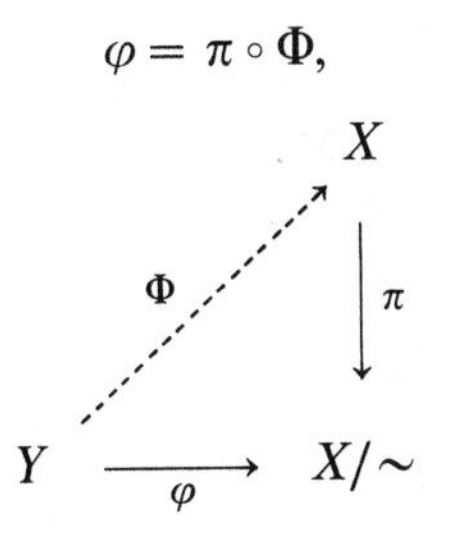

or even if this is the case only locally, i.e. if every $y \in Y$ has a neighborhood U for which there is a continuous map $\Phi_U: U \to X$ with $\pi \circ \Phi_U = \varphi | U$,

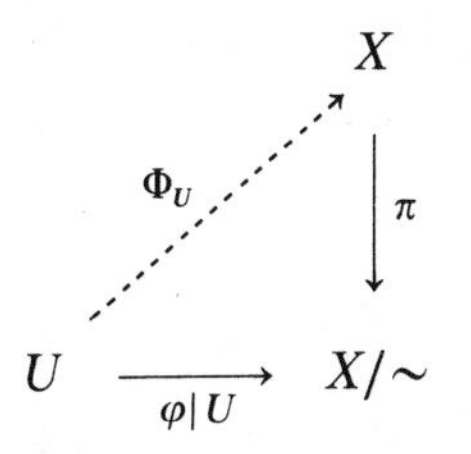

then φ is of course continuous.

§3. Properties of Quotient Spaces

Which properties of X are carried over to $X/\sim$? Connectedness and compactness are the most well-behaved ones:

Note. *If X is (path-)connected (resp. compact), then so is $X/\sim$ (being a continuous image of X).*

The situation is entirely different as regards the third property discussed in Chapter I: A quotient space of a Hausdorff space is in general not Hausdorff anymore. A trivial reason why this may be so is that the equivalence classes may not be all closed:

Note. *A necessary condition for a quotient space $X/\sim$ to be Hausdorff is that all equivalence classes in X be closed; for if $y \notin [x]$ is a boundary point of $[x]$, then $[x]$ and $[y]$ cannot be separated by disjoint neighborhoods in $X/\sim$.*

Another, possibly more elegant way to say this is the following: The closedness of equivalence classes in X is equivalent to the closedness of points in $X/\sim$, and of course in a Hausdorff space all points are closed sets.

All right, so let the equivalence classes be closed: this condition is perfectly reasonable, and without it nothing works. But then what? Here are two misleadingly similar examples. In both $X = \mathbb{R}^2$ with the usual topology, the equivalence classes are closed one-dimensional submanifolds, arranged in a very simple way, and the decomposition of $\mathbb{R}^2$ is even invariant under translations in the y direction. Moreover, the two examples are so similar that one cannot easily describe the difference between them by means of point-set theoretical properties of the two equivalence relations—except that one of the quotient spaces is Hausdorff and the other is not!

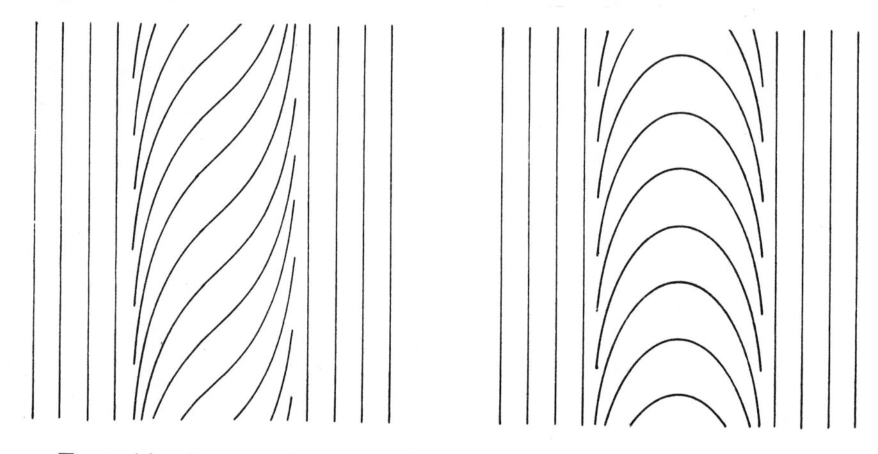

From this what we can learn right away is that the separation properties of quotients depend a lot on the particular arrangement of the equivalence classes, and one should be thankful for the existence of theorems that guarantee Hausdorffness for whole classes of examples.

§4. Examples: Homogeneous Spaces

Let's recall the following notations from algebra: If G is a group and $H \subset G$ is a subgroup, then G/H denotes the set $\{gH \mid g \in G\}$ of left cosets of H, which are the equivalence classes under the equivalence relation on G defined by $a \sim b \Leftrightarrow b^{-1}a \in H$. If moreover H is a normal subgroup (i.e. $gHg^{-1} = H$ for all $g \in G$), then G/H inherits a canonical group structure from G.

Definition (Topological Group). A group G that is also a topological space is called a *topological group* if $G \times G \to G$, $(a, b) \mapsto ab^{-1}$ is continuous.

The groups $\mathrm{GL}(n, \mathbb{R})$ and $\mathrm{GL}(n, \mathbb{C})$ of invertible $n \times n$ matrices, for example, are topological groups in a canonical way, and so are the abelian groups $(E, +)$ underlying topological vector spaces. Also, any subgroup of a topological group, endowed with the subspace topology, is trivially a topological group.

Definition (Homogeneous Space). If $H \subset G$ is a subgroup of a topological group G, the quotient space G/H is called a homogeneous space.

Thus our general definition of quotient space $X/\sim$ in §1 is applied here for the case $X = G$ and $a \sim b :\Leftrightarrow b^{-1}a \in H$.

Why are homogeneous spaces of interest? This is a damned far-reaching question and cannot be fully answered at the level of this book. But I will try to make some comments about it. When topological groups are found in nature, they are generally not given abstractly as a set G with a composition law and a topology, but concretely, as a group of transformations, i.e. a group of bijective maps from a set X onto itself, and the law of composition is none other than the composition of maps. But such an X should not be imagined as being just a set, any more than G should be the group of *all* bijections of X. The set X will instead be endowed with some more structure: first a topology, but maybe also a differentiable or analytical or algebraic or metrical or linear or any other sort of structure, according to the situation. And the elements $g: X \to X$ of G will be bijections that are compatible with that structure. From this connection it generally follows what topology we must reasonably endow G with. Take $G = \mathrm{GL}(n, \mathbb{R})$ as a simple example: The set X will be $\mathbb{R}^n$ with its linear structure.

So far this is an observation about topological groups and not yet about homogeneous spaces. But now think of some mathematical object A in X or on X or somehow associated with X and its structure: for instance, a given subset $A \subset X$ or a function $A: X \to \mathbb{C}$, or indeed anything such that it makes sense to say that A is transformed by $g \in G$ into a similar object gA, and gA is transformed by $h \in G$ into $(hg)A$. For a subset $A \subset X$, gA is simply the image $g(A)$; for a function $A: X \to \mathbb{C}$, gA is the function $A \circ g^{-1}: X \to \mathbb{C}$, and so on. But now the set $H = \{g \in G \mid gA = A\}$ of elements that transform A into itself is a subgroup of G, and *the homogeneous space G/H can be considered in a natural way to be the space of all positions that A can assume under transformations by elements of G.* As a simple example of this process consider $G = O(n + k)$ and $X = \mathbb{R}^k \times \mathbb{R}^n$. Let A be the subspace $\mathbb{R}^k \times 0$. The orthogonal matrices that take $\mathbb{R}^k \times 0$ into itself are exactly those of the form

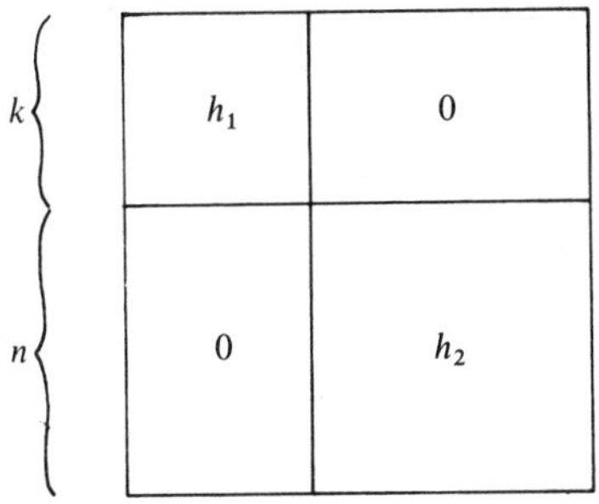

$$\begin{array}{r} k\left\{ \vphantom{h_1} \right. \\ n\left\{ \vphantom{h_2} \right. \end{array} \begin{pmatrix} h_1 & 0 \\ 0 & h_2 \end{pmatrix}$$

with $h_1 \in O(k)$ and $h_2 \in O(n)$; we have thus $H = O(k) \times O(n) \subset O(n + k)$, and the homogeneous space $O(n + k)/O(k) \times O(n)$ is the so-called "Grassmannian manifold" of k-dimensional subspaces of $\mathbb{R}^{n+k}$. In the case $k = 1$, for instance, $O(n + 1)/O(1) \times O(n)$ is the well-known real projective space $\mathbb{RP}^n$ of straight lines through the origin of $\mathbb{R}^{n+1}$.

Returning now to the general case, it often happens that it is such a "position space" that we are primarily interested in, and discovering groups G and H such that it can be represented as a homogeneous space G/H will be the first step in studying it.

With this I have described, vaguely enough, a first important point of view from which homogeneous spaces are interesting; a second, related one (homogeneous spaces as "orbits"), will be discussed in the next section; and now I will mention a third one, quite profound. In a very general way, one of the fundamental principles in the study of complicated geometrical objects is to decompose them into simpler parts and study the laws according to which the whole can be reconstructed from the parts. One such possibility is the decomposition of a space into similar "fibers". Now the rules according to which such similar fibers can be reassembled into "fiber bundles" are determined by a topological group, the "structure group", and in connection with these topological groups homogeneous spaces come to the fore again. For instance, the Grassmannian manifolds $O(n + k)/O(k) \times O(n)$ are important for the classification of vector bundles, and the knowledge we acquire about these homogeneous spaces (the Grassmannian manifolds) pays off as a means to analyze vector bundles, which in turn .... But this is taking us too far afield. Let me just say one more thing: Apart from being instrumental in achieving any immediate ends, as outlined above, homogeneous spaces deserve attention in themselves as geometrical objects, being both very varied and, as group quotients, accessible to the methods of the theory of topological groups (or other groups with richer structure). Cf. the "symmetric spaces" of Riemannian geometry.

*

All this takes us way beyond the scope of point-set topology: my modest purpose of convincing you of the real occurrence of homogeneous spaces in mathematics may have been achieved, and we can gradually set our feet back on the ground.

To conclude, let's turn again to the question of Hausdorffness of quotient spaces. Intractable as this may be in the general case, for homogeneous spaces a very neat criterion holds:

Lemma (for the proof, which in any case is not difficult, see for instance Bourbaki [2], III.12). *A homogeneous space G/H is Hausdorff if and only if H is closed in G.*

If E is a topological vector space and $E_0 \subset E$ is a subspace, the quotient E/E_0, with the quotient topology, is again a topological vector space. Now since the closure $\bar{E}_0$ of E_0 is also a vector subspace of E, the above lemma implies that $E/\bar{E}_0$ is always a Hausdorff space, and in particular we call $E/\overline{\{0\}}$ the Hausdorff space *associated with E*. For instance, if the topology of E is given by a seminorm $|..|$, we have $\overline{\{0\}} = \{x \in E \,|\, |x| = 0\}$, and $|..|$ defines a norm on $E/\overline{\{0\}}$.

§5. Examples: Orbit Spaces

Definition. Let G be a topological group and X a topological space. A continuous action or operation of G on X is defined to be a continuous map $G \times X \to X$, denoted by $(g, x) \mapsto gx$, such that:

Axiom 1. $1x = x$ for all $x \in X$.
Axiom 2. $g_1(g_2 x) = (g_1 g_2)x$ for all $x \in X$ and $g_1, g_2 \in G$.

Thus every g defines via $x \mapsto gx$ a map from X into itself, and the two axioms mean that this correspondence is a group homomorphism from G into the group of bijections of X onto itself. Because of the continuity of $G \times X \to X$ the image of this homomorphism is actually contained in the group of homeomorphisms of X into itself.

Definition (G-space). A G-space is a pair consisting of a topological space X and a continuous G-action on X.

Differentiable G-manifolds are defined analogously: G is then not only a topological group, but actually a Lie group (i.e. G is a differentiable manifold and $G \times G \to G$, $(a, b) \mapsto ab^{-1}$ is differentiable), X is not only a topological space but actually a differentiable manifold, and the action $G \times X \to X$ is not only continuous but differentiable.

G-spaces and especially G-manifolds are the object of an extensive theory, the *theory of transformation groups*. To be sure, we can't go deeper into this theory here, but there is only one tiny aspect of it which concerns us in connection with this chapter: namely, that the quotient topology plays a role in it from the very first concepts on, as I'll presently explain.

Definition (Orbit). If X is a G-space and $x \in X$, the set $Gx := \{gx \,|\, g \in G\}$ is called the orbit or trajectory of x.

The orbit is thus the set of points into which x can be taken by the action of elements of the group. In particular, if G is the additive group $(\mathbb{R}, +)$ of the real numbers, a G-action is the same as a "flow" (cf. the theory or ordinary differential equations, integration of vector fields), the orbits are the images of integral curves or trajectories of the flow, and the name "trajectory" is generalized to any group by analogy.

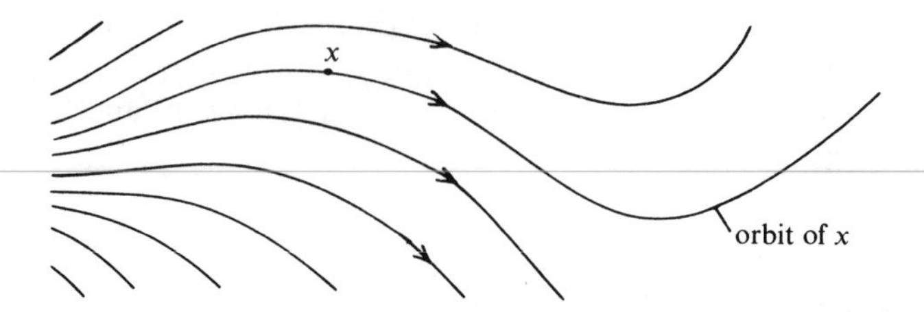

The orbits are the equivalence classes of the equivalence relation defined by "$x \sim y :\Leftrightarrow y = gx$ for some $g \in G$", and we can thus consider the quotient topology on the set of orbits.

Definition (Oribt Space). If X is a G-space, the set of orbits, endowed with the quotient topology, is called the orbit space and denoted by X/G.

As an illustration we will "calculate" the orbit space for a simple example: this means we are going to find a homeomorphism between the orbit space and a well-known topological space. Let $G = \mathrm{SO}(2)$, the group of rotations of $\mathbb{R}^2$ around the origin, let X be the unit sphere $S^2 = \{x \in \mathbb{R}^3 \mid \|x\| = 1\}$, and let the G-action on X be given by rotation around the x_3-axis, i.e.

$$g(x_1, x_2, x_3) := (g(x_1, x_2), x_3).$$

The orbits are then the parallels of latitude and the two poles.

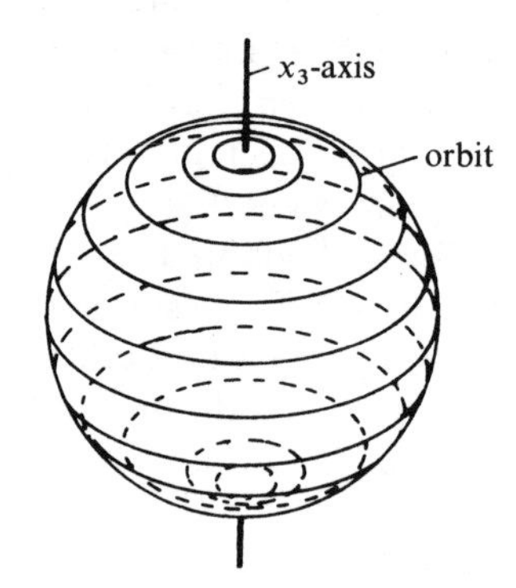

Assertion. $S^2/G \cong [-1, 1]$.

PROOF. Consider the continuous map $\pi_3 : S^2 \to [-1, 1]$ given by projection on the x_3-axis. Since π_3 is constant on each orbit, it defines a map

$$f_3 : S^2/G \to [-1, 1]$$

such that the diagram

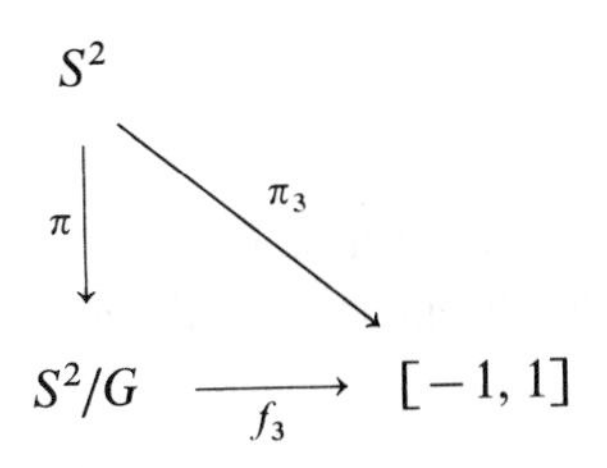

commutes; moreover, f_3 is obviously bijective. By §2, f_3 is continuous as well; but S^2/G is compact, since it is the continuous image of the compact set S^2, and $[-1, 1]$ is Hausdorff. Hence f_3 is a homeomorphism by the theorem at the end of Chapter I, qed. □

To conclude we will take a look at the individual orbits themselves, and discover a quotient topology in them as well.

Definition (Stabilizer). Let X be a G-space and $x \in X$. Then

$$G_x := \{g \in G \mid gx = x\}$$

is called the stabilizer or isotropy group of the point x.

Remark. The correspondence $gG_x \mapsto gx$ defines a continuous bijection from the homogeneous space G/G_x onto the orbit Gx.

PROOF. First, $gG_x \mapsto gx$ really gives rise to a well-defined map $G/G_x \to Gx$, because $gG_x = hG_x$ implies $h = ga$ for some $a \in G_x$ and hence $hx = gax = gx$. This map is obviously surjective, and also injective since $gx = hx$ implies $h^{-1}gx = x$, hence $h^{-1}g \in G_x$ and $hG_x = gG_x$. Continuity follows from §2, because the composition $G \to G/G_x \to Gx$ is continuous. qed. □

This is already a quite close connection between orbits and homogeneous spaces. Now when in particular G is compact and X is Hausdorff, G/G_x is compact since it is the continuous image of G. Then Gx is Hausdorff because it is a subspace of a Hausdorff space, and we obtain from our theorem at the end of Chapter I that $G/G_x \to Gx$ is a homeomorphism: then the orbits really "are" homogeneous spaces.

§6. Examples: Collapsing a Subspace to a Point

So far we have considered examples of quotient topologies which come up "spontaneously", as it were, in mathematics, by conferring the obvious topology to an object already given in some way. In §§6 and 7 we are introduced to the quotient operation more as a handicraft technique, used according to one's aims and purposes to manufacture new topological spaces with given properties.

Definition. Let X be a topological space, $A \subset X$ a non-empty subspace. Denote by X/A the quotient space $X/\sim_A$ obtained from the equivalence relation defined by

$$x \sim_A y :\Leftrightarrow x = y \quad \text{or } x, y \text{ both belong to } A.$$

The equivalence classes are thus A and the one-point sets not contained in A; in the quotient space X/A, therefore, A is a point, whereas the complement $X \setminus A$ remains unaltered. Incidentally, this explains why it has been found convenient to set $X/\varnothing := X + \{pt\}$ in the case $A = \varnothing$. The process of going from X to X/A is called collapsing of A into a point.

Analogously, one can of course collapse several subspaces to points, and we'll introduce a notation for this process:

Definition. If X is a topological space and $A_1, \ldots, A_r \subset X$ are disjoint non-empty subsets, denote by $X/A_1, \ldots, A_r$ the quotient space obtained from the equivalence relation defined by

$$x \sim y :\Leftrightarrow x = y \text{ or there is an } i \text{ such that } x, y \text{ both belong to } A_i.$$

Remark. As already noticed in §3, $X/A_1, \ldots, A_r$ can only be Hausdorff if the A_i are all closed. In "reasonable" spaces this condition is in fact sufficient as well, e.g. it is not difficult to prove that $X/A_1, \ldots, A_r$ is a Hausdorff space if the A_i are closed and X is *metrizable*. Of course it is essential here that there be only a finite number of equivalence classes with more than one point; otherwise we've already got a counterexample in §3.

Example 1 (Cone Over a Set). Let X be a topological space. Then

$$CX := X \times [0, 1]/X \times 1$$

is called the cone over X.

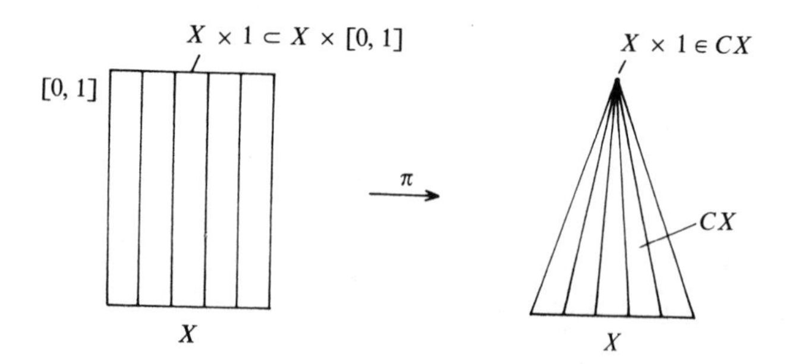

Now again, such a picture is to be understood only as a mental image, but is it well chosen as an image? Shouldn't we, since the complement $X\backslash A$ remains unaltered when we build X/A, represent the cone like this:

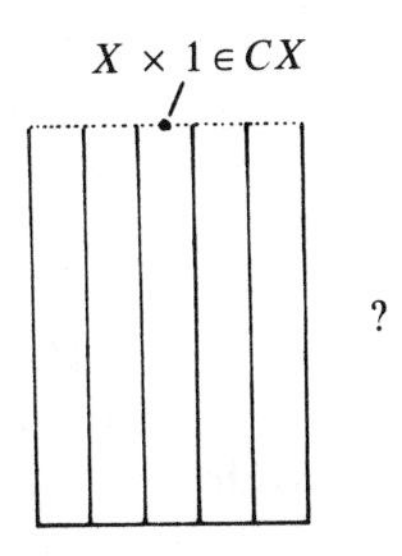

No, this picture creates a false image of the topology of the cone, because by definition of the quotient topology every neighborhood of the vertex of the cone must have as inverse image a neighborhood of the "lid" $X \times 1$ of the cylinder, which is the case only in the first figure.

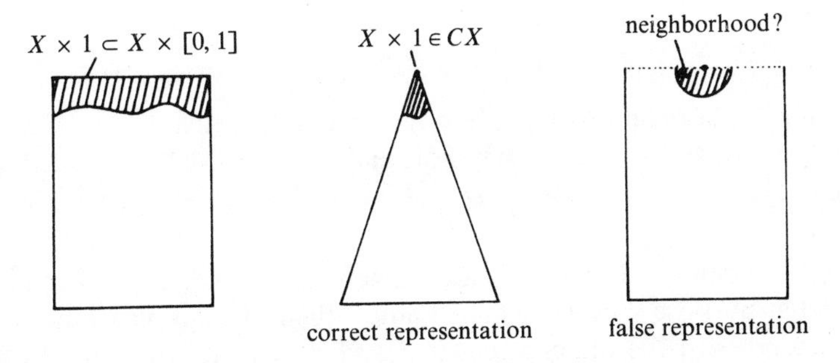

And the fact that nothing happens with the complement of the lid is also adequately taken into account by the first picture, in that it shows that the canonical projection π gives rise to a homeomorphism from $X \times [0, 1]\backslash X \times 1$ onto $CX\backslash\{X \times 1\}$.

Example 2 (Suspension). For a topological space X, the space

$$\Sigma X := X \times [-1, 1]/X \times \{-1\}, X \times \{1\}$$

is called the suspension of X or double cone over X:

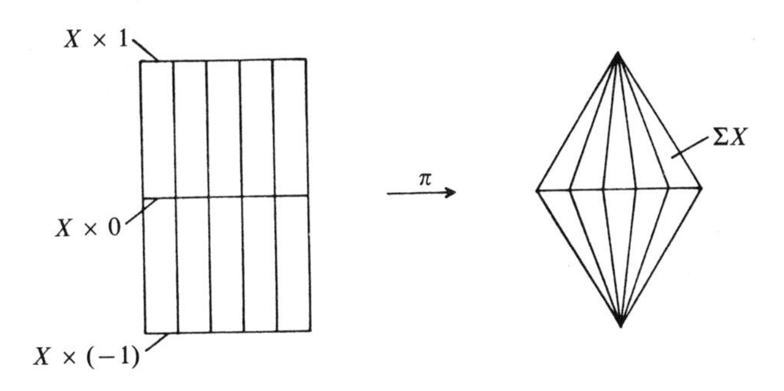

Example 3. Sometimes there is reason to construct a cone over only a part of X, but keeping the whole of X as the "ground". Thus, if $A \subset X$, denote by $C_A X$ the quotient $(X \times 0 \cup A \times [0, 1])/A \times 1$:

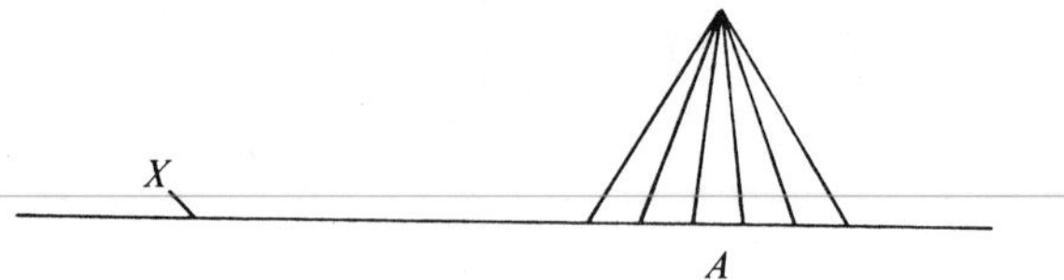

Example 4 (Wedge and Smash Products). Let X and Y be topological spaces and $x_0 \in X$, $y_0 \in Y$ fixed points. Then denote by $X \vee Y$ (wedge product) the subspace $X \times y_0 \cup x_0 \times Y$ of the product $X \times Y$,

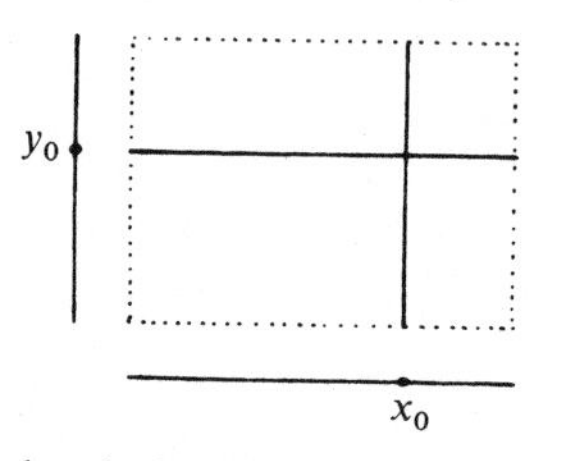

and by $X \wedge Y$ (smash product) the quotient space $X \times Y/X \vee Y$.

Example 5 (Thom Space). Let E be a vector bundle with a Riemannian metric, $DE := \{x \in E \mid \|v\| \leq 1\}$ its disc bundle and $SE := \{v \in E \mid \|v\| = 1\}$ its sphere bundle. Then the quotient space DE/SE is called the Thom space of the bundle E.

All these constructions come up in algebraic topology, but I can't explain now what purpose they serve, and I admit the last example cannot even be understood with the material covered so far—I included it just to have it "in reserve". However, let's take a closer look at the simplest case of this example, namely the case when E has only one "fiber": $E = \mathbb{R}^n$. Then DE is the closed ball D^n and SE is the sphere S^{n-1}. What is the outcome when we collapse the whole boundary of the ball to a point? So? It is a space homeomorphic to the n-sphere S^n. In fact, pick a continuous map $f: D^n \to S^n$ that takes the boundary S^{n-1} onto the south pole p and that takes $D^n \setminus S^{n-1}$ bijectively onto $S^n \setminus p$ (one example of such a map is obtained by mapping the radii in the obvious way onto the great semicircles ("meridians") that run from the north to the south pole).

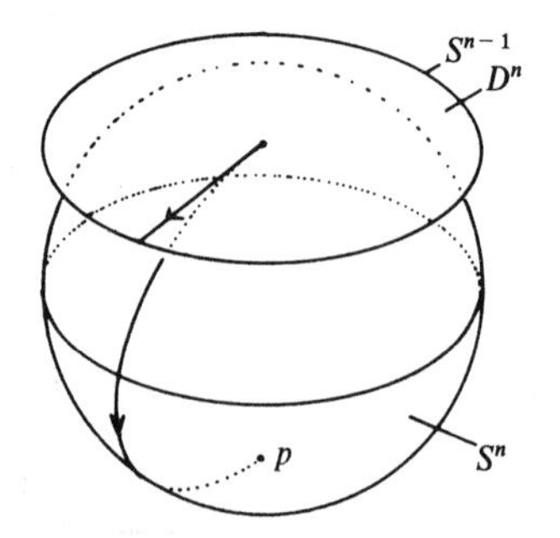

Then one obtains via f a bijection $\varphi: D^n/S^{n-1} \to S^n$, with $f = \varphi \circ \pi$:

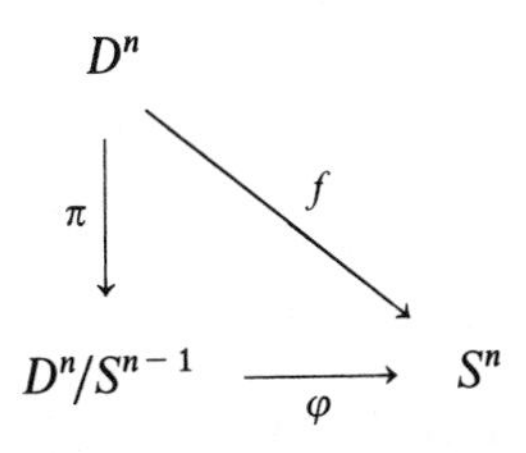

From §2 we know φ is continuous, and thus in fact a homeomorphism because it is a bijection from the compact space D^n/S^{n-1} onto the Hausdorff space S^n.

§7. Examples: Gluing Topological Spaces Together

Definition. Let X and Y be topological spaces, $X_0 \subset X$ a subspace and $\varphi: X_0 \to Y$ a continuous map. Then denote by $Y \cup_\varphi X$ the quotient space $X + Y/\sim$ by the equivalence relation on $X + Y$ generated by $x \sim \varphi(x)$ for all $x \in X_0$. One also says that $Y \cup_\varphi X$ is obtained by *attaching* or *gluing* X to Y by means of the *attaching map* φ, or by *identifying* points $x \in X_0$ with their images $\varphi(x) \in Y$.

Just in case, let me repeat it again in detail: The equivalence classes either have one point (every point of $X + Y$ that does not belong to either X_0 or $\varphi(X_0)$), or have the form $\varphi^{-1}(y) + \{y\} \subset X + Y$.

Example 1. Let X be a topological space and $\varphi: S^{n-1} \to X$ be continuous. One says then that $X \cup_\varphi D^n$ is obtained from X by "attaching a cell" by means of the attaching map φ.

(We'll discuss cell attaching again in Chapter VII—"CW-complexes".)

What is the relationship between the "building blocks" X and Y and the space $X \cup_\varphi Y$? Since no two different points of Y are identified with one another, Y is always contained in $Y \cup_\varphi X$ in a canonical way, or, more exactly, the canonical map $Y \subset X + Y \to Y \cup_\varphi X$ is injective, and there will be no misunderstanding in writing $Y \subset Y \cup_\varphi X$. And all the more reason to do so since the subspace topology inherited by $Y \subset Y \cup_\varphi X$ coincides with the original topology on Y, as one can easily verify (you'll have to use the continuity of φ). Thus:

Note. *Y is a subspace of $Y \cup_\varphi X$ in a canonical way.*

This is, of course, not true for the attached part X: The complement $X \backslash X_0$ is a subspace of $Y \cup_\varphi X$, but X itself can be considerably changed by the canonical continuous map $X \subset X + Y \to Y \cup_\varphi X$. For instance, if Y is a single point, then $\{pt\} \cup_\varphi X$ is exactly the X/X_0 defined in §6.

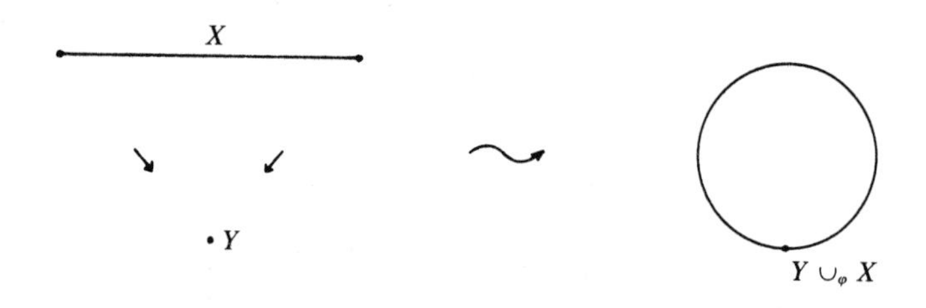

If, however, φ is a homeomorphism from X_0 onto a subspace $Y_0 \subset X$ and $\psi: Y_0 \to X_0$ is its inverse, we of course have $Y \cup_\varphi X = X \cup_\psi Y$, and by the note above both spaces X and Y are canonically contained in $Y \cup_\varphi X$. The following examples are of this type:

Example 2. Attaching a "handle" $D^k \times D^{n-k}$ to an n-dimensional manifold with boundary, by means of an embedding $\varphi: S^{k-1} \times D^{n-k} \to \partial M$, as is done in Morse theory (see, for example, [14]):

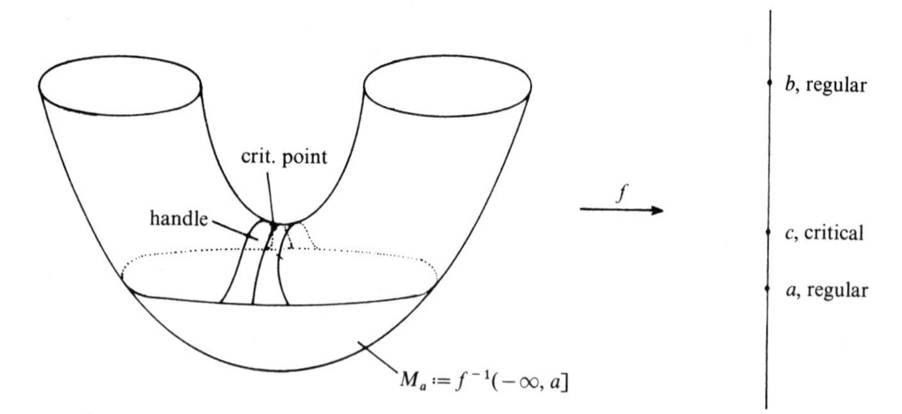

Writing $M_y := f^{-1}(-\infty, y]$ we have that crossing a "critical point" is essentially equivalent to attaching a handle: $M_b \cong M_a \cup_\varphi (D^k \times D^{n-k})$.

Example 3. In differential topology one constructs the so-called "connected sum" $M_1 \# M_2 := (M_2 \backslash p_2) \cup_\varphi (M_1 \backslash p_1)$ of two manifolds (see for instance [3], p. 102), where one first "punctures" the manifolds, i.e. takes away an arbitrary point, and then glues them together via an appropriate φ.

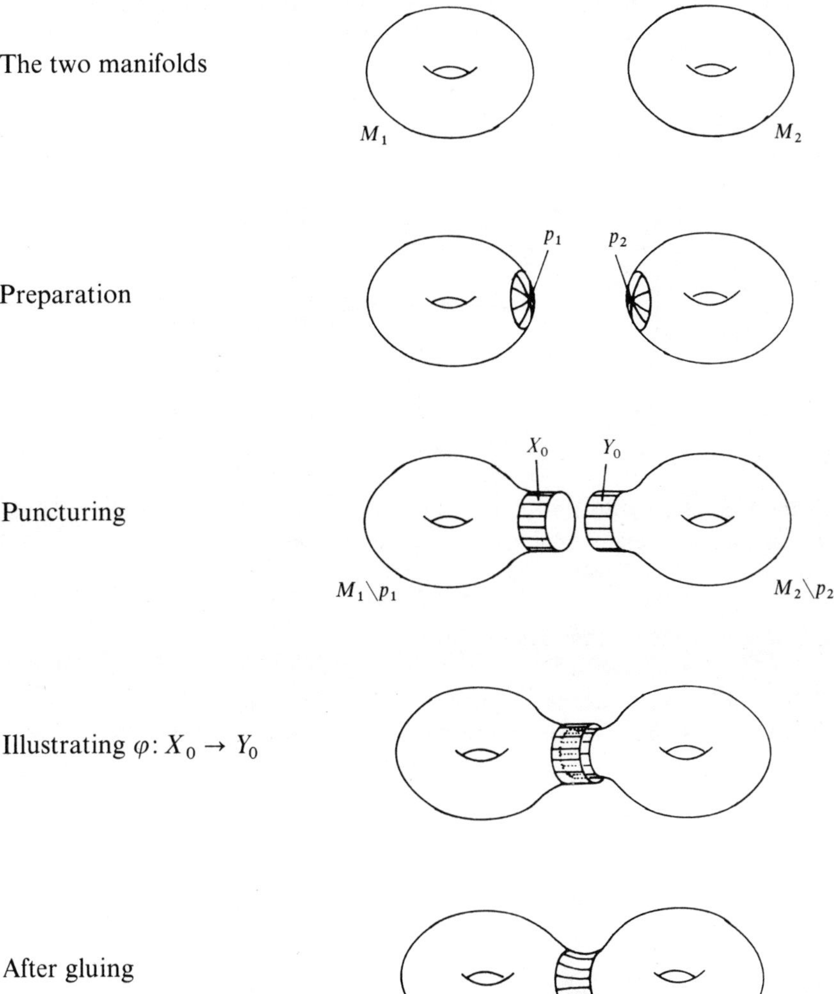

Till now we've always been gluing two spaces X and Y, that is, forming a quotient space $X + Y/\sim$. One can similarly glue a space X to itself in many different ways, by "identifying" certain points of X to others by prescribing some map, which is the same as introducing an equivalence relation and passing to the quotient $X/\sim$. We'll introduce the following notation having in mind the two examples below:

Notation. Let X be a topological space and $\alpha: X \to X$ a homeomorphism. Then denote by $X \times [0, 1]/\alpha$ the quotient space of $X \times [0, 1]$ by the equivalence relation given by $(x, 0) \sim (\alpha(x), 1)$, which in particular means that all other points (x, t), $0 < t < 1$, are equivalent only to themselves.

Example 4 (Möbius Strip). If $X = [-1, 1]$ and $\alpha(x) := -x$, then $X \times [0, 1]/\alpha$ is homeomorphic to the Möbius strip

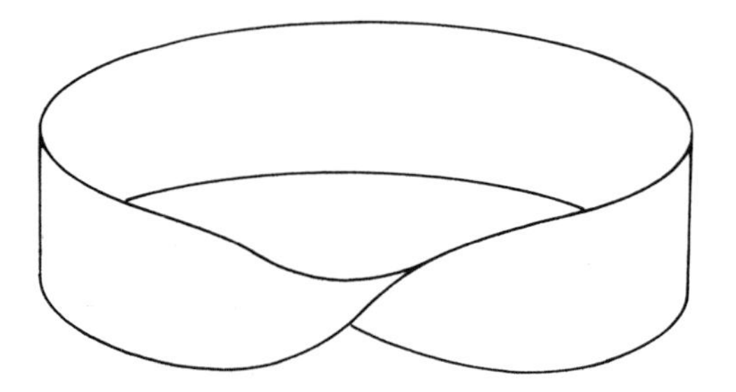

Example 5 (Klein Bottle). Let $\alpha: S^1 \to S^1$ denote reflection on the x-axis, that is, $\alpha(z) := \bar{z}$, where S^1 is seen as the circle $\{z \in \mathbb{C} \mid |z| = 1\}$. Then

$$S^1 \times [0, 1]/\alpha$$

is homeomorphic to the "Klein bottle".

It is not so easy to visualize the Klein bottle concretely, as there is no subspace of $\mathbb{R}^3$ homeomorphic to it. To get an idea of what it "looks like" one has to resort to the trick of "apparent intersection". To illustrate this trick consider the following figure:

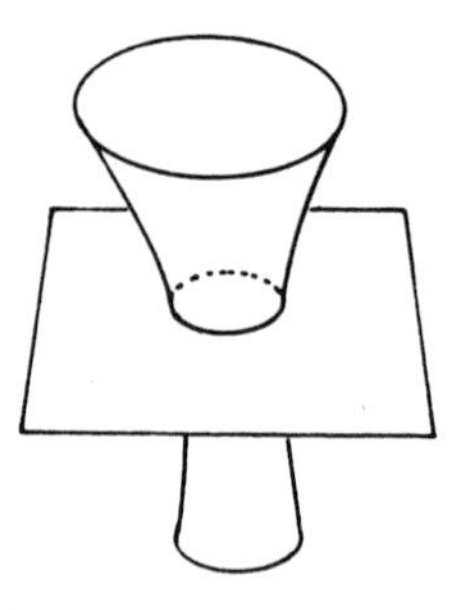

One would normally interpret it as representing a subset of $\mathbb{R}^3$, the union of a rectangle and a funnel which intersect in a circle. But if we are requested to consider the intersection of the two parts as being merely *apparent*, the figure acquires an altogether different meaning. It no longer stands for that subset of $\mathbb{R}^3$, but becomes an admittedly imperfect attempt to represent a space which is the topological, and thus disjoint, sum of the rectangle and the funnel. Thus in this space the circle is present twice: once in the rectangle and once in the funnel; and we no longer have the possibility, originally suggested by the figure alone, of passing continuously from the funnel to the rectangle.

If you want to think of it in a more concrete way, imagine the space in question as a subspace of $\mathbb{R}^4$, and the figure representing its projection on $\mathbb{R}^3 \times 0$. The rectangle could be wholly contained in $\mathbb{R}^3 \times 0$, and for the funnel also the invisible fourth coordinate would be mostly zero, except around the apparent intersection, where it would be positive. The following two-dimensional analogy can clarify this situation:

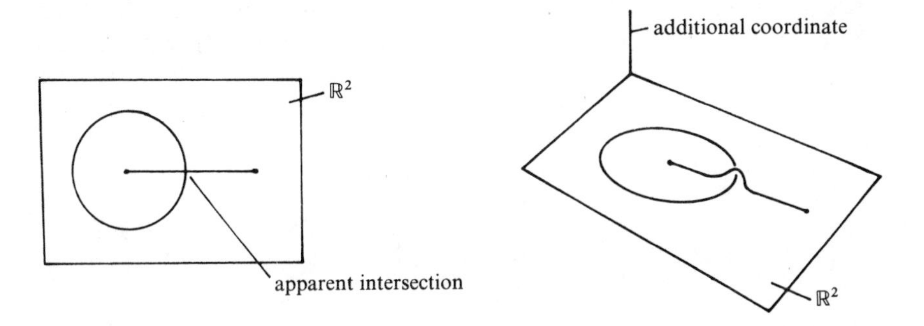

But when you really use a figure with apparent intersections to understand some property of a given space, you'll notice that this fourth-dimensional crutch is not necessary at all and all you have to do is be prepared to keep the two parts mentally separate at the apparent intersection.

In this spirit we can thus visualize a cylinder $S^1 \times [0, 1]$ with an apparent self-intersection as in (6):

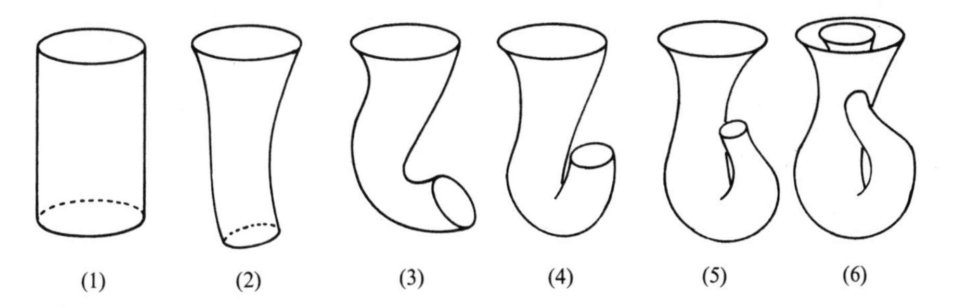

The movement suggested by the series (1)–(6) indicates what point of (1) corresponds to what point in (6). Except for translation and reduction, the "bottom" $S^1 \times 0$ of (1) is tipped over exactly once around the axis that goes through the points $(1, 0)$ and $(-1, 0)$, in the transition from (1) to (6). Hence in (6) we have each pair of points to be identified, $(z, 0)$ and $(\bar{z}, 1)$, exactly in

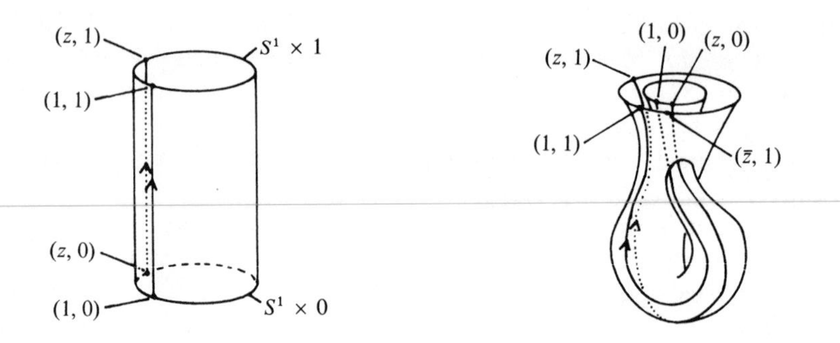

front of each other, and all we have to do is enlarge the inner boundary circle a little bit to realize spatially the identification that defines the Klein bottle. Except that we don't want the seam to look like a sharp edge (there is no reason why it should be visually distinguishable from the other "parallels of latitude" of the figure), and therefore we'll do it in the following way:

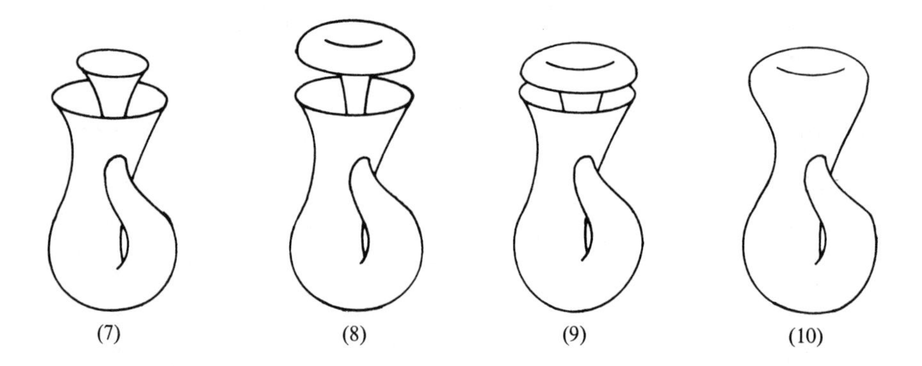

Then what we get in (10) is the Klein bottle, represented with an apparent self-intersection.

If we now cut in two this curious contraption, to find out what it looks like inside,

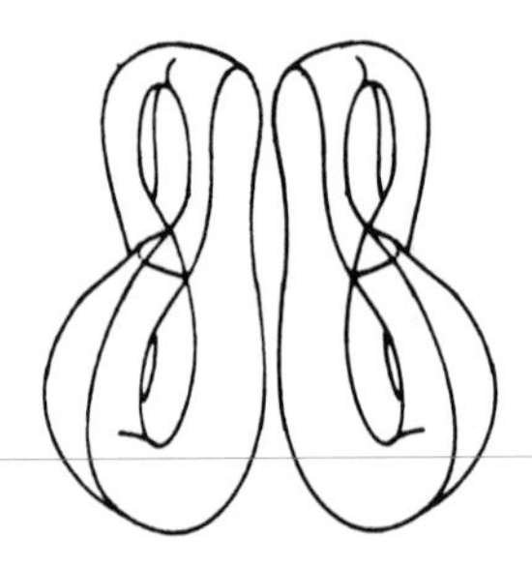

and subsequently undo the apparent intersection very carefully in both halves, we obtain, after smoothing and flattening things a bit, two Möbius strips:

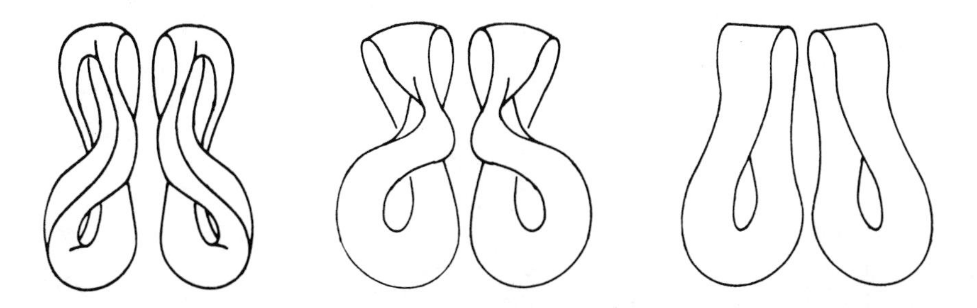

Again, if we follow this process backwards, we see that two Möbius strips M glued together along the boundaries become a Klein bottle:

$$M \cup_{\mathrm{Id}_{\partial M}} M \cong K.$$

Hm ... Have we proved this now? In no way. A *proof* would have to look like this: First define a map

$$([-1, 1] \times [0, 1]) + ([-1, 1] \times [0, 1]) \to S \times [0, 1]$$

by writing $(\theta, t) \mapsto (e^{\pi i\theta/2}, t)$ on the first and $(-e^{-\pi i\theta/2}, t)$ on the second summand; then, prove that this really gives rise to a well defined bijection $M \cup_{\mathrm{Id}_{\partial M}} M \to K$; then prove the continuity of this bijection using the note in §2, and finally apply the theorem at the end of Chapter I that says that a continuous bijection from a compact into a Hausdorff space is always a homeomorphism.

*

It is often said against intuitive, spatial argumentation that it is not really argumentation but just so much gesticulation—just "handwaving". Shall we then abandon all intuitive arguments? Certainly not. As long as it is backed by the gold standard of rigorous proofs, the paper money of gestures is an invaluable aid for quick communication and fast circulation of ideas. Long live handwaving!

CHAPTER IV

Completion of Metric Spaces

§1. The Completion of a Metric Space

In this chapter things really depend on the metric of the metric spaces and not only on the topology given by the metric, but it is customary and it makes sense to rank metric spaces among the objects of point-set topology, and we will not be pedantic about such distinctions anyway.

Let's recall that a sequence $(x_n)_{n \geq 1}$ in a metric space (X, d) is called a *Cauchy sequence* if for every $\varepsilon > 0$ there is an n_0 such that $d(x_n, x_m) < \varepsilon$ whenever $n, m \geq n_0$. The metric space (X, d) is said to be *complete* if every Cauchy sequence converges.

For instance, the real line $\mathbb{R}$ with the usual metric $d(x, y) := |x - y|$ is complete, as every mathematics student soon finds out (completeness axiom for the real numbers); thus $\mathbb{R}^n$ with the usual metric is also complete; Hilbert and Banach spaces are complete by definition; every compact metric space is complete—and finally we can obtain an enormous number of other examples simply by observing that a subspace of a complete metric space, that is a subset $A \subset X$ with the metric $d | A \times A$, is complete if and only if A is closed in X.

Now by completion is understood the process of making a non-complete space (X, d) into a complete one $(\hat{X}, \hat{d})$ by adjoining as few new points as possible:

Definition (Completion). Let (X, d) be a metric space. A metric space $(\hat{X}, \hat{d})$ such that $X \subset \hat{X}$ and $d = \hat{d} | X \times X$ is called a *completion* of (X, d) if

1. $(\hat{X}, \hat{d})$ is complete, and
2. X is dense in $\hat{X}$, i.e. the closure $\overline{X}$ of X in $\hat{X}$ is equal to $\hat{X}$ itself.

The second condition says exactly that $\hat{X}$ is a *minimal* complete space such that X is a subspace of $\hat{X}$: Since X is dense, each "new" point $x \in \hat{X} \setminus X$ is the limit of a sequence $(x_n)_{n \geq 1}$ of points in X, and if we took $\hat{x}$ away, $(x_n)_{n \geq 1}$ would become a non-convergent Cauchy sequence and completeness would no longer hold. Can every metric space be completed, and if so, in what different ways? A good rule of thumb in such situations is to deal first with the uniqueness question, and here the latter is easily settled by the following

Proposition (Uniqueness of the Completion). *If $(\hat{X}, \hat{d})$ and $(\tilde{X}, \tilde{d})$ are completions of a metric space (X, d), there is exactly one isometry $\hat{X} \xrightarrow{\cong} \tilde{X}$ whose restriction to X is the identity.*

Proof. The image of $x = \lim x_n$ under such an isometry, where (x_n) is a Cauchy sequence in X, must of course be the limit $\tilde{x}$ of the same sequence in $\tilde{X}$, which exists by assumption; thus there is at most one such isometry. Conversely, if (x_n) and (y_n) are Cauchy sequences in X and $\hat{x}, \hat{y}$ (resp. $\tilde{x}, \tilde{y}$) are their limits in $\hat{X}$ (resp. $\tilde{X}$), we have $\hat{d}(\hat{x}, \hat{y}) = \lim d(x_n, y_n) = \tilde{d}(\tilde{x}, \tilde{y})$; hence a map $\hat{X} \to \tilde{X}$ given by $\hat{x} \mapsto \tilde{x}$ is first of all well-defined, and, second, has the required property of being an isometry such that $x \mapsto x$ for all $x \in X$, qed. □

In this sense ("up to a canonical isometry") there is at most one completion of (X, d), and for this reason it is for most purposes irrelevant *how* we construct it, if we can do it in the first place. It is easy to find a completion in a natural way when X is already a metric subspace of a complete metric space Y: all we need to do is take the closure of X in Y. In the examples below, $Y = \mathbb{R}^2$ and X is in each case a subspace homeomorphic to $\mathbb{R}$:

Example 1. $X = \mathbb{R}$, complete

Example 2. X is an open half-line; completed with a point

Example 3. X is an open interval; completed with two points

Example 4.

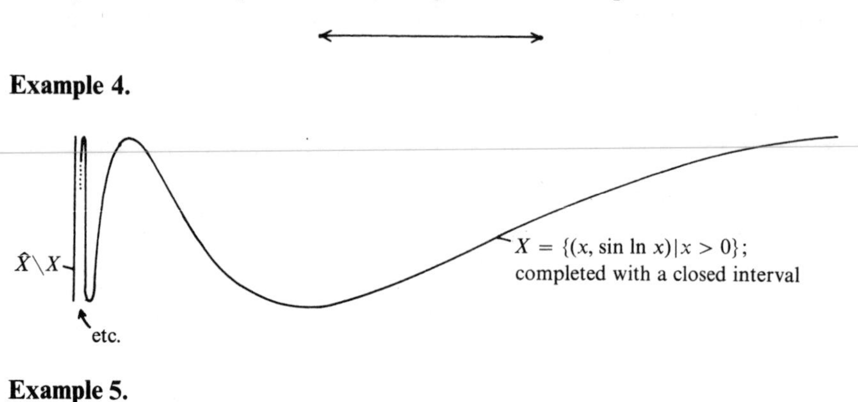

Example 5.

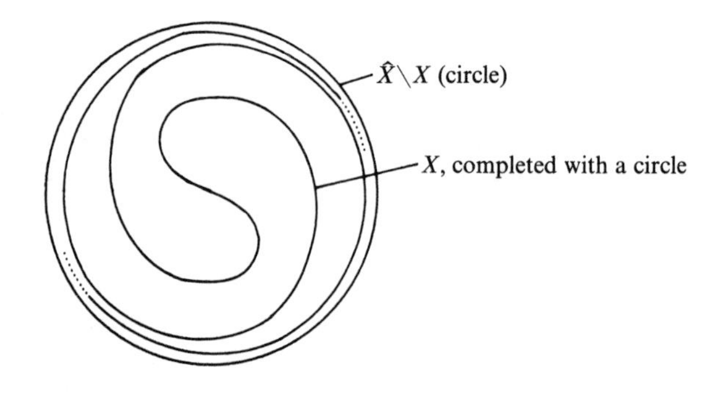

These examples are meant to show right away that homeomorphic metric spaces can have extremely non-homeomorphic completions.

Let us now tackle the problem of constructing a completion for an arbitrary metric space (X, d). We obviously have to create new points to be limit values of the non-convergent Cauchy sequences ("ideal" points, as they were formerly called, implying that, strictly speaking, they do not exist). And moreover two non-convergent Cauchy sequences $(a_n)_{n \geq 1}$ and $(b_n)_{n \geq 1}$ have the same "ideal limit point" $\hat{x}$ if and only if $\lim_{n \to \infty} d(a_n, b_n) = 0$, because then and only then they are to have the same limit in a completion of X.

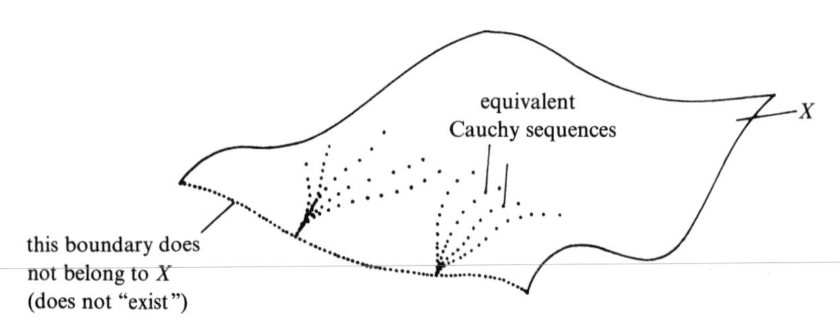

Now where do we get these points $\hat{x}$ from? Well, this is one such case in which we can make use of the "paradisiacal" (à la Hilbert) possibilities of Cantor's set theory: We simply take, as ideal limit points corresponding to an equivalence class of non-convergent Cauchy sequences, the equivalence class itself!

Lemma (Existence of the Completion). *Let (X, d) be a metric space and $\mathcal{N}$ the set of non-convergent Cauchy sequences in X. Two Cauchy sequences (a_n) and (b_n) will be called equivalent if $\lim_{n\to\infty} d(a_n, b_n) = 0$. Defining now the set $\hat{X}$ as the disjoint union $X + \mathcal{N}/\sim$ and the distance d on X by*

$$\hat{d}(x, y) := d(x, y),$$

$$\hat{d}(x, a) = \hat{d}(a, x) := \lim_{n\to\infty} d(a_n, x),$$

$$\hat{d}(a, b) = \lim_{n\to\infty} d(a_n, b_n),$$

for all x, y in X and all equivalence classes $a = [(a_n)]$ and $b = [(b_n)]$ in $\mathcal{N}/\sim$, we obtain a well-defined map $\hat{d}: \hat{X} \times \hat{X} \to \mathbb{R}$ such that $(\hat{X}, \hat{d})$ is a completion of (X, d).

Proof. . . . The proof consists of that sort of argument that does not become any clearer by being detailed by somebody else. One proves successively that $\hat{d}$ is well-defined, that $\hat{d}$ satisfies the axioms for a metric, then that X is dense in $\hat{X}$ and that $(\hat{X}, \hat{d})$ is complete. Just in case, let me indicate, for the proof of completeness, that the terms of a Cauchy sequence $(\hat{x}_n)$ don't have to be all in X. Choose sequences $(x_{nk})_{n\geq 1}$ in X such that either $[(x_{nk})_{k\geq 1}] = \hat{x}_n$ or, if $\hat{x}_n \in X$, $x_{nk} = \hat{x}_n$ for all k. Then $(x_{nk_n})_{n\geq 1}$, for an appropriate sequence $k_1 < k_2 < \cdots$, is a Cauchy sequence, and $(\hat{x}_n)$ will converge towards its limit. . . . qed. □

We all, reader and author, value this trick with the equivalence classes of non-convergent Cauchy sequences for what it is, a formal vehicle, and yet none of us will lay aside his intuition and start imagining the ideal limit points as really being such bouquets of Cauchy sequences. But look at what can be done with set theory in the wrong hands and especially in school... Well, never mind.

To conclude, a short side note about a question of formulation and presentation. We could have defined the completion, perhaps more elegantly, in the following way: Let $\mathscr{C}$ be the set of *all* (convergent and non-convergent) Cauchy sequences in X. Let $\hat{X} = \mathscr{C}/\sim$ and $\hat{d}([a_n], [b_n]) = \lim d(a_n, b_n)$. Then $(\hat{X}, \hat{d})$ is a complete metric space, and if X is "looked on" as a subset $X \subset \hat{X}$ via $x \to [(x)_{n\geq 1}]$, then $(\hat{X}, \hat{d})$ is a completion of (X, d). (Proof: . . . .) Indeed, in analogous situations, this version is often preferred. Who would introduce the field of complex numbers as $\mathbb{C} := \mathbb{R} \cup \{(x, y) \in \mathbb{R}^2 \mid y \neq 0\}$, endowed with such and such composition laws? One naturally sets $\mathbb{C} := \mathbb{R}^2$ as a set etc.

and then announces that in the future $\mathbb{R}$ is "to be considered" as a subset of $\mathbb{C}$ via $x \mapsto (x, 0)$. And yet I must admit I always feel rotten when I have to make this announcement to beginners...

§2. Completion of a Map

Let (X, d) be a metric space and $f: X \to Y$ a continuous map. Under what circumstances and how can one extend f to a continuous map $\hat{f}: \hat{X} \to Y$?

First a preliminary remark. There is at most one way to do this:

Proposition. *Let A be a topological space, $X \subset A$ a dense subset, i.e. $\overline{X} = A$, and let $f, g: A \to B$ be two continuous maps in a Hausdorff space B that coincide on X. Then $f = g$.*

PROOF. If f and g are different at a point $a \in A$, then they are different in a whole neighborhood $f^{-1}(U) \cap g^{-1}(V)$ of a, hence $a \notin \overline{X}$, contradiction. qed. □

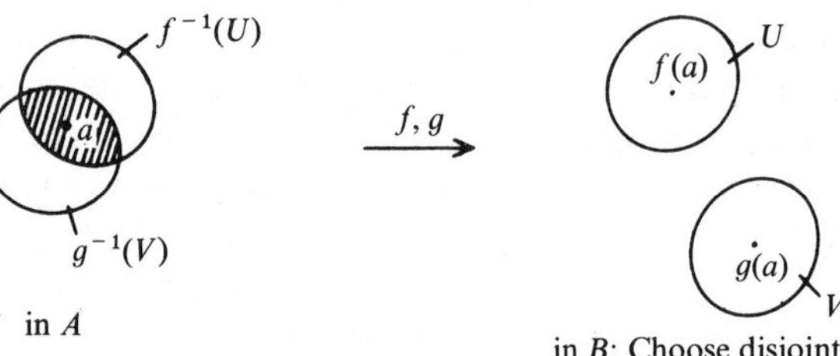

in A

in B: Choose disjoint neighborhoods U, V of $f(a)$, $g(a)$...

Thus in particular a continuous map from a metric space X into a Hausdorff space can be extended to the completion $\hat{X}$ in at most one way. But sometimes it cannot be done at all, and there are in fact two different types of obstacles, as illustrated by the following examples:

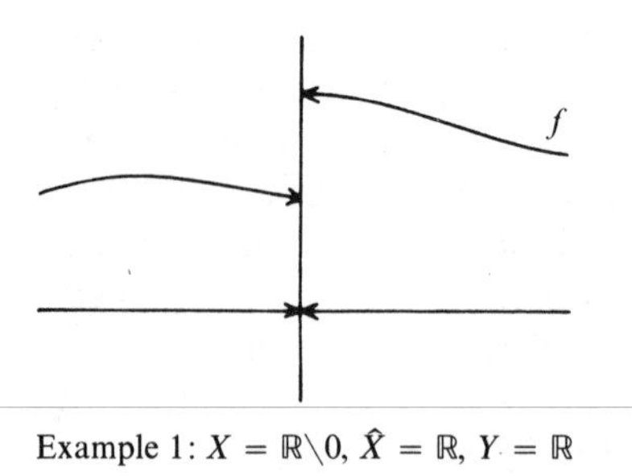

Example 1: $X = \mathbb{R}\backslash 0$, $\hat{X} = \mathbb{R}$, $Y = \mathbb{R}$

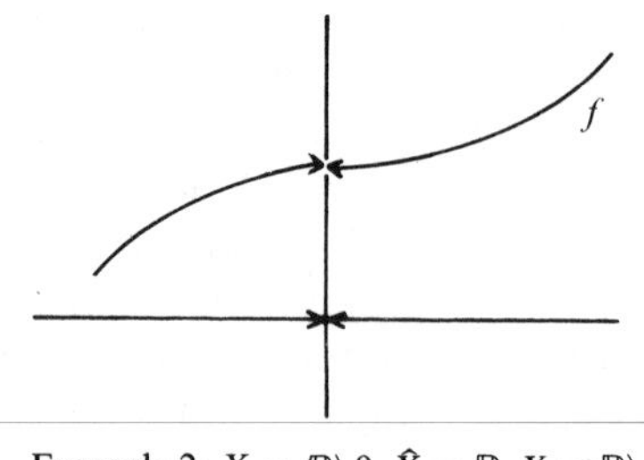

Example 2: $X = \mathbb{R}\backslash 0$, $\hat{X} = \mathbb{R}$, $Y = \mathbb{R}\backslash 1$

In the first case f has a "jump" at point 0, and therefore cannot be extended continuously. In the second case f doesn't have any jumps, but the only value that would do for a continuous extension is "missing" from the counter-domain. We'll now suppose that Y is also a metric space and smooth away these two difficulties by using appropriate hypotheses: To guarantee that the counterdomain doesn't have any "holes", we simply take its completion: and to avoid jumps of f at the ideal limit points, we take f to be uniformly continuous.

Let's recall that if (X, d) and (Y, d') are metric spaces, a map $f: X \to Y$ is called *uniformly continuous* if for every $\varepsilon > 0$ there is a $\delta > 0$ such that $d'(f(a), f(b)) < \varepsilon$ for *all* $a, b \in X$ with $d(a, b) < \delta$.

Lemma (Completion of Maps). *Let (X, d) and (Y, d') be metric spaces and $f: X \to Y$ a uniformly continuous map. Then if $(\hat{X}, \hat{d})$ and $(\hat{Y}, \hat{d}')$ are completions of (X, d) and (Y, d'), there is exactly one extension of f to a continuous map $\hat{f}: \hat{X} \to \hat{Y}$.*

Proof. Because of uniform continuity Cauchy sequences are taken into Cauchy sequences, and their equivalence is preserved. Hence defining $\hat{f}(\lim_{n\to\infty} x) := \lim_{n\to\infty} f(x_n)$ we obtain a well-defined extension of f to a map $\hat{f}: \hat{X} \to \hat{Y}$, where $(x_n)_{n \geq 1}$ denotes a non-convergent Cauchy sequence in X and the limits refer to convergence in $\hat{X}$ and $\hat{Y}$ respectively. One easily verifies that f is continuous, even uniformly continuous.... qed. □

In particular let us notice en passant that isometries are always uniformly continuous ($\delta = \varepsilon$), and the completion $f: \hat{X} \to \hat{Y}$ of an isometry $f: X \xrightarrow{\cong} Y$ is of course again an isometry.

§3. Completion of Normed Spaces

It is not surprising that the concept of "completeness" is particularly important for the function spaces of analysis, since interesting functions, "solutions" to whatever it may be, are often constructed as limits of function sequences. As already mentioned in Chapter II, §5, one can talk about "Cauchy sequences", hence completeness and incompleteness, in arbitrary topological vector spaces; and when we try to establish axiomatically the fundaments for these concepts in topological spaces, we are led to the notion of "uniform spaces": a structure situated between metric and topology (every metric space being in particular a uniform space, and every uniform space a topological space), and we can carry through the completion of uniform spaces in the same way as with metric spaces. Every topological vector space is also a uniform space, in a canonical way. But here I will limit myself to normed topological vector spaces.

First a couple of easy-to-prove notes of a general nature: The completion of a normed space $(E, \|..\|)$ is a Banach space $(\hat{E}, \|..\|\hat{})$ in a canonical way: The vector space structure on $\hat{E}$ can be elegantly defined as the quotient of the vector space of all Cauchy sequences in E by the vector subspace of all sequences that approach 0. The norm $\|..\|: E \to \mathbb{R}$ is uniformly continuous ($\varepsilon = \delta$), and can thus be continuously extended to $\|..\|\hat{}: \hat{E} \to \mathbb{R}$, which is again a norm (check this!), and $\hat{d}(x, y) = \|x - y\|\hat{}$. The completion of a real or complex inner product space is a Hilbert space in a canonical way.

Continuous linear maps $f: E \to V$ between normed spaces are automatically uniformly continuous, and the extensions $\hat{f}: \hat{E} \to \hat{V}$ obtained as described above are again linear.

Now to get closer to what I'm really driving at, let's recall (or define) the meaning of "L^p spaces": For a fixed $p \geq 1$ let $\mathscr{L}^p(\mathbb{R}^n)$ denote the vector space of the Lebesgue-measurable functions $f: \mathbb{R}^n \to \mathbb{R}$ for which $|f|^p$ is Lebesgue-integrable. Then

$$\|f\|_p := \sqrt[p]{\int_{\mathbb{R}^n} |f|^p \, dx}$$

gives a seminorm on $\mathscr{L}^p(\mathbb{R}^n)$. As we know from integration theory, the closure of point 0, that is $\overline{\{0\}} = \{f \in \mathscr{L}^p \mid \|f\|_p = 0\}$, is exactly the set of all functions that vanish outside some set of measure zero. $L^p(\mathbb{R}^n)$ will be defined as the corresponding normed space: $L^p(\mathbb{R}^n) := \mathscr{L}^p(\mathbb{R}^n)/\overline{\{0\}}$ (see end of III, §4). Then an important theorem of integration theory asserts that $L^p(\mathbb{R}^n)$ is complete, hence a Banach space.

Analogously one defines the L^p space $L^p(X, \mu)$ for an arbitrary measure space $(X, \mathfrak{M})$ with a σ-additive measure $\mu: \mathfrak{M} \to [0, \infty]$ etc. The case $p = 2$ is particularly nice, because $L^2(X, \mu)$ is even a Hilbert space with

$$\langle f, g \rangle = \int_X fg \, d\mu.$$

Upon closer inspection an L^p-space is thus a fairly intricate mathematical object, and those who didn't study Lebesgue integrals because they thought they could get along with Riemann integrals have a well-founded horror of such spaces. But $L^p(\mathbb{R}^n)$ contains also very harmless elements, and in particular the vector space $C_0^\infty(\mathbb{R}^n)$ of infinitely often differentiable functions $f: \mathbb{R}^n \to \mathbb{R}$ with "compact support" (i.e. vanishing outside a compact set),

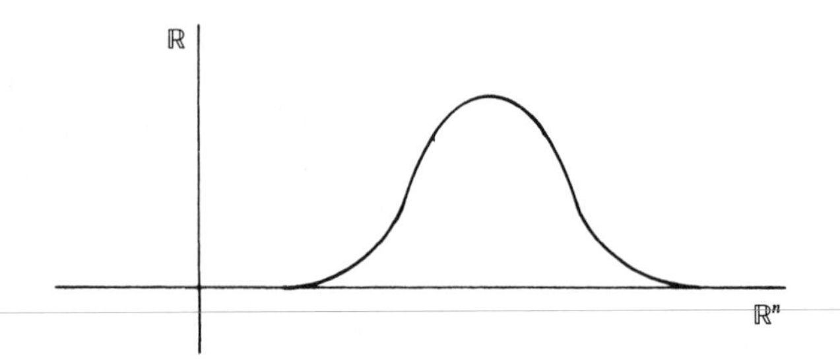

is a subspace of $L^p(\mathbb{R}^n)$ in a canonical way. On this subspace the p-norm

$$\|f\|_p = \sqrt[p]{\int_{\mathbb{R}^n} |f|^p \, dx}$$

and the scalar product $\langle f, g\rangle = \int_{\mathbb{R}^n} fg \, dx$ are very easily understood, even with a rudimentary knowledge of any of the concepts of integral. It is thus encouraging to know that integration theory asserts that $C_0^\infty(\mathbb{R}^n)$ *is dense in* $L^p(\mathbb{R}^n)$—because this then means that $L^p(\mathbb{R}^n)$ is a completion of

$$(C_0^\infty(\mathbb{R}^n), \|\,..\,\|_p),$$

and since a space has only one completion up to a canonical isometric isomorphism, we can *define* $L^p(\mathbb{R}^n)$ as the completion of $(C_0^\infty(\mathbb{R}^n), \|\,..\,\|_p)$ (or $(C_0^\infty(\mathbb{R}^n), \langle .., ..\rangle)$ if $p = 2$)!

Now I don't want to generate the illusion that with this simple completion trick one can really get the Lebesgue integral out of the way, because if L^p spaces are introduced as above as completions, we don't know anything about to what extent those new "ideal limit points" can be regarded as functions, or how they can be otherwise interpreted in an analytical manageable way. But still! The fact that $C_0^\infty(\mathbb{R}^n)$ and like spaces, together with any tailor-made norms they may be given according to the problem at hand, can be completed like this—zap—creates an invaluable freedom of movements (irrespective of the necessity of studying the ideal points). An example to conclude the chapter will make clear what I'm trying to say.

To denote *partial differential operators* it is customary to use multi-index notation: If $\alpha = (\alpha_1, \ldots, \alpha_n)$, where $\alpha_i > 0$ are integers and $|\alpha| := \alpha_1 + \cdots + \alpha_n$, the symbol D^α means $\partial^{|\alpha|}/\partial x_1^{\alpha_1} \cdots \partial x_n^{\alpha_n}$; this is thus the general form of a partial derivative of any order $|\alpha|$. Now let $a_\alpha\colon \mathbb{R}^n \to \mathbb{R}$ be functions (let's say smooth). Then $p = \sum_{|\alpha| \le k} a_\alpha D^\alpha$ is a linear partial differential operator on $\mathbb{R}^n$, and an equation of the form $Pf = g$, where g is given on $\mathbb{R}^n$ and f is to be determined, is called a (non-homogeneous) linear partial differential equation.

I have intentionally left unsaid what the "operator" P "operates" on. In any case P defines a linear map $P\colon C_0^\infty(\mathbb{R}^n) \to C_0^\infty(\mathbb{R}^n)$ in the linear algebra sense. But linear algebra by itself won't get us any further; it would be much better if we could regard P as a continuous operator in a Hilbert space, for instance, because then we'd be able to resort to the function-analytical theory of such operators in our study of P.

Now we can of course complete $C_0^\infty(\mathbb{R}^n)$ turning it into the Hilbert space $L^2(\mathbb{R}^n)$, but unfortunately there is no way P can operate on it: $P\colon C_0^\infty \to C_0^\infty$ itself is not $\|\,..\,\|_2$-continuous, much less a possible extension $\hat{P}\colon L^2 \to L^2$. But there are many other ways of defining scalar products on C_0^∞, to be chosen according to the ends in view (which is admittedly easier said than done). The most obvious ones are perhaps the scalar products given on $C_0^\infty(\mathbb{R}^n)$ by

$$\langle f, g\rangle_r := \sum_{|\alpha| \leq r} \int_{\mathbb{R}^n} \langle D^\alpha f, D^\alpha g\rangle \, dx$$

for every integer $r \geq 0$. The Hilbert spaces $H^r(\mathbb{R}^n)$ obtained by completing $(C_0^\infty(\mathbb{R}^n), \langle .., ..\rangle_r)$ are the simplest examples of objects called "Sobolev spaces", which are used as a quite refined tool in the theory of partial differential operators. It is not difficult to see that $P = \sum_{|\alpha|<k} a_\alpha D^\alpha$, with appropriate conditions on the coefficients, defines a linear operator

$$P^r: H^r(\mathbb{R}^n) \to H^{r-k}(\mathbb{R}^n)$$

which is in fact continuous.

This example illustrates what sort of use point-set topology has in analysis. Of course the study of the differential operator P is by no means completed with the introduction of Sobolev spaces, and topology cannot solve the analytical problems proper, but it creates a climate in which analysis thrives.

CHAPTER V

Homotopy

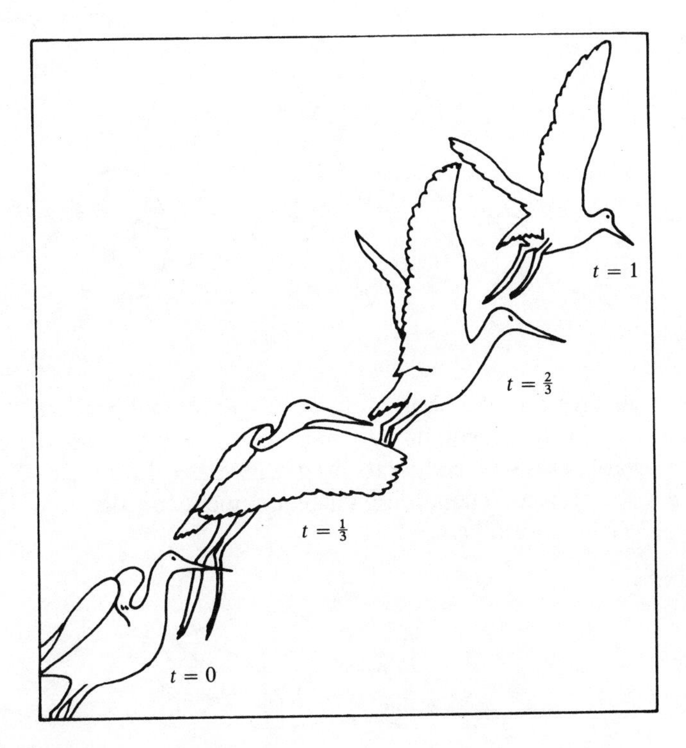

§1. Homotopic Maps

In §1–3 I will define and intuitively explain the basic notions of "homotopic maps", "homotopy" and "homotopy equivalence", and in §4–7 we discuss the use of these notions.

Definition (Homotopy, Homotopic). Two continuous maps $f, g: X \to Y$ between topological spaces are called *homotopic*, $f \simeq g$, if there is a *homotopy* h between them, i.e. a continuous map $h: X \times [0, 1] \to Y$ with $h(x, 0) = f(x)$ and $h(x, 1) = g(x)$ for all $x \in X$.

Notations. In this case we also write $f \underset{h}{\simeq} g$. We denote by $h_t: X \to Y$, for t fixed, the continuous map given by

$$h_t(x) := h(x, t).$$

Then we have $h_0 = f$ and $h_1 = g$.

To the extent that maps can graphically visualized in the first place, so can homotopies: Think of $[0, 1]$ as a time interval; at time $t = 0$ the map h_t

has the form f, but it changes in the course of time until it takes the form g at time $t = 1$. The whole change must go on continuously in both variables, so one can also say that the homotopy h is a "continuous deformation of f into g".

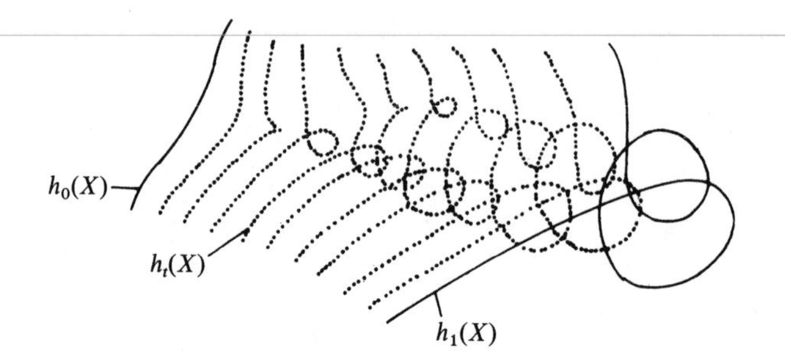

One frequently considers homotopies with additional properties besides continuity; you will be acquainted from function theory with the concept of homotopy of paths with endpoints fixed, where $X = [0, 1]$, $Y \subset \mathbb{C}$ is open and $p, q \in Y$ are fixed. The additional requirements on the homotopy are $h_t(0) = p$, $h_t(1) = q$ for all t:

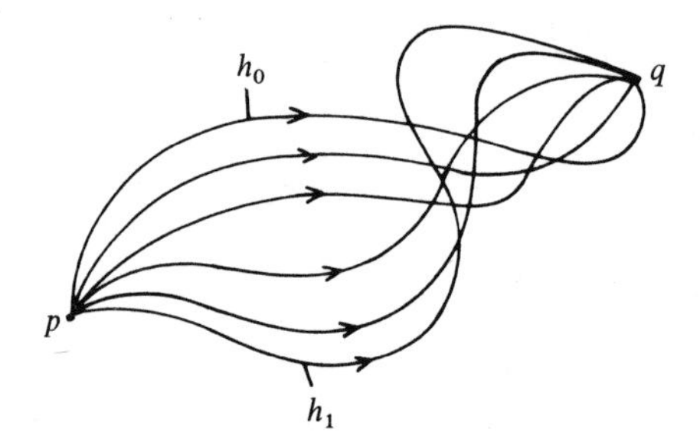

In differential topology one often considers homotopies between maps from one manifold to another, such that each h_t is required to be an embedding (h is an "isotopy") or a diffeomorphism (h is a "diffeotopy"); thus there are many situations in which h has to satisfy one or another additional requirement. But here we're just going to discuss the basic notion, where h has to be merely continuous.

As already hinted by the sign $\simeq$, "homotopic" is an equivalence relation. Reflexivity is clear: $f \simeq f$ because $h_t := f$ for all t gives a homotopy between f and f. For symmetry, if $f \simeq g$ via h_t, $0 \leq t \leq 1$, then $g \simeq f$ via h_{1-t}. For transitivity, if $f \underset{h}{\simeq} g \underset{k}{\simeq} l$, we have $f \underset{H}{\simeq} l$ with

$$H_t = \begin{cases} h_{2t} & \text{for } 0 \leq t \leq \frac{1}{2} \\ k_{2t-1} & \text{for } \frac{1}{2} \leq t \leq 1 \end{cases} \qquad \text{(prove continuity!)}$$

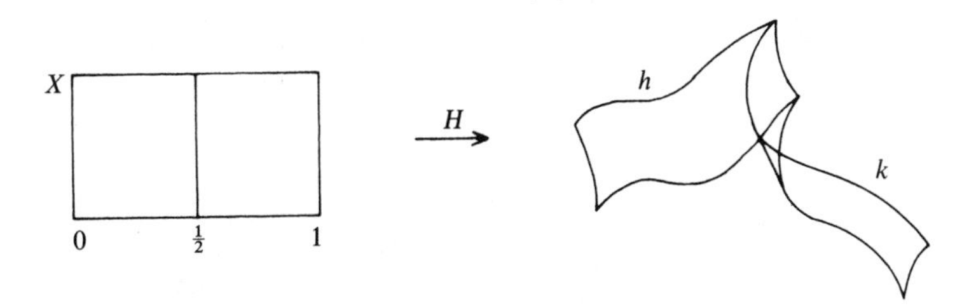

Notation. If X and Y are topological spaces, $[X, Y]$ denotes the set of equivalence classes ("homotopy classes") of continuous maps of X into Y.

If we introduce an $(n + 1)$-th concept, we can naturally formulate its relation to the n previous ones by means of n lemmas, for instance:

Note (Composition of Homotopic Maps). *For two pairs of homotopic maps* $X \underset{f \simeq g}{\longrightarrow} Y \underset{\bar{f} \simeq \bar{g}}{\longrightarrow} Z$, *the compositions* $\bar{f} \circ f$ *and* $\bar{g} \circ g$ *are also homotopic (via* $\bar{h}_t \circ h_t$ *etc.)*

Note (Product of Homotopic Maps). *For two pairs of homotopic maps* $f_i \simeq g_i \colon X_i \to Y_i$, $i = 1, 2$, *the maps* $f_1 \times f_2$ *and* $g_1 \times g_2$ *from* $X_1 \times X_2$ *into* $Y_1 \times Y_2$ *are also homotopic (via* $h_t^{(1)} \times h_t^{(2)}$ *etc.)*

But for such simple concept as homotopy it would be pedantic to present now a list of such results as complete as possible; let's wait leisurely to see what we'll really need in the applications.

§2. Homotopy Equivalence

Definition (Homotopy Equivalence). A continuous map $f \colon X \to Y$ is called a *homotopy equivalence* between X and Y if it possesses a "homotopy inverse", i.e. a continuous map $g \colon Y \to X$ with $g \circ f \simeq \mathrm{Id}_X$ and $f \circ g \simeq \mathrm{Id}_Y$.

We then say that f and g are homotopy equivalences inverse to each other, and call the spaces X and Y homotopy equivalent. Compositions of homotopy equivalences are evidently homotopy equivalences, and so is of course Id_X for all X; thus we really have an equivalence relation: $X \simeq X$, and $X \simeq Y \simeq Z$ implies $Y \simeq X$ (anyway) and $X \simeq Z$.

A simple but important special case:

Definition (Contractible Space). A topological space is called *contractible* if it is equivalent to a space with one point.

In this case the definition reduces to the requirement that there be a homotopy $h \colon X \times [0, 1] \to X$, called a "contraction", between the identity and a constant map $X \to \{x_0\} \subset X$. The space $\mathbb{R}^n$, for instance, is contractible, for $h_t(x) := (1 - t)x$ defines a contraction into the origin; and so is of course any star-shaped subspace of $\mathbb{R}^n$.

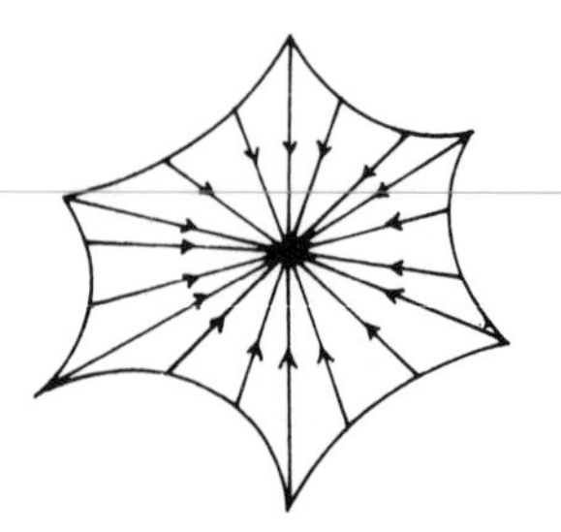

Definition (Retract and Deformation retract). Let X be a topological space, $A \subset X$ a subspace. A is called a *retract* of X if there is a *retraction* $\rho: X \to A$, i.e. a continuous map with $\rho | A = \text{Id}_A$. Now if ρ is also homotopic to the identity, as a map from X into X, then ρ is called a *deformation retraction* and A, correspondingly, a *deformation retract*. Finally, if this homotopy between ρ and Id_X can be chosen so that all points of A are kept fixed in the course of it, i.e. if $h_t(a) = a$ for all $t \in [0, 1]$ and all $a \in A$, then ρ is called a *strong* deformation retraction and A a *strong* deformation retract of X.

A deformation retraction $\rho: X \to A$ and the corresponding inclusion $i_A: A \subset X$ are evidently homotopy equivalences inverse to each other, because "retract" already means that $\rho \circ i_A = \text{Id}_A$, and "deformation" implies $i_A \circ \rho \simeq \text{Id}_X$.

Now this whole thing about retracts may taste at first rather dry, but I will presently say something that will whet your interest in strong deformation retracts. When dealing with homotopies in practice, it is important to develop an eye for homotopy equivalences of spaces. Whenever possible, one tries to avoid having to laboriously look for an $f: X \to Y$ and a $g: X \to Y$ and a homotopy $f \circ g \simeq \text{Id}_Y$ and another one $g \circ f \simeq \text{Id}_X$, and writing all this stuff down in tiresome detail. We want to be able to say at a glance: These two spaces are homotopy equivalent, and everybody shall agree: Of course, they are homotopy equivalent.

But this quick identification of homotopy equivalent spaces is in many practical cases based on the construction of the homotopy equivalence $X \simeq Y$ as a composition $X \simeq X_1 \simeq \cdots \simeq X_r \simeq Y$, where in each step (there are generally only a few of them, sometimes only one) the two spaces are either homeomorphic or one is a strong deformation retract of the other. Why are strong deformation retracts so easy to recognize, though? Well, what does such a deformation do: It takes each point of X along a continuous path into the space A, between times 0 and 1, and all we have to mind is that points that already start in A do not move. So if X and A can be graphically depicted at all, the deformation h, if it exists, will probably be easy to find. Now let's consider some examples to help sharpening our eye.

§3. Examples

Example 1. The origin is of course a strong deformation retract of $\mathbb{R}^n$ or of the ball D^n.

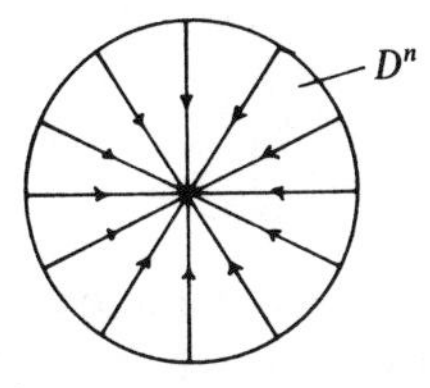

But this means $A \times 0$, where A is any topological space, is a strong deformation retract of $A \times \mathbb{R}^n$ or $A \times D^n$,

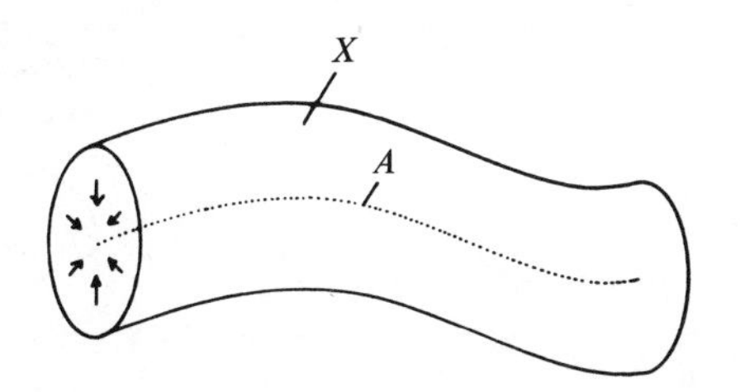

and in particular the solid torus $S^1 \times D^2$, for instance, is homotopy equivalent to the circle S^1:

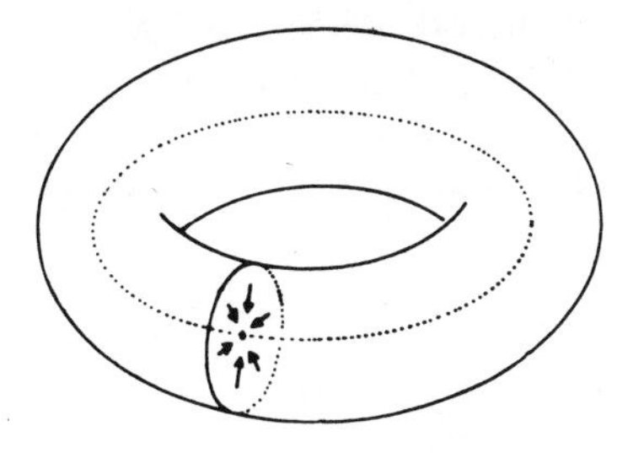

More generally, if E is a vector bundle over a topological space A, the zero section is a strong deformation retract of E, or, if E is endowed with a Riemannian metric, of the disc bundle DE, i.e. $A \simeq E \simeq DE$.

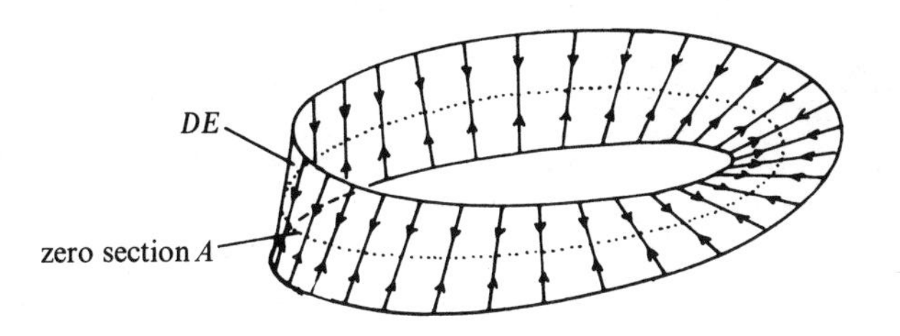

Example 2. The sphere $S^{n-1} = \{x \in \mathbb{R}^n \mid \|x\| = 1\}$ is a strong deformation retract of the ball minus a point $D^n \backslash 0$:

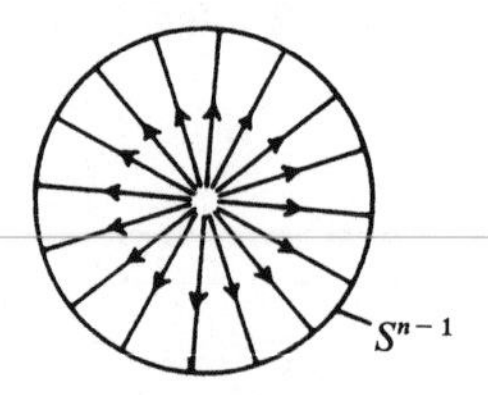

If we attach a cell to a space X and remove the origin from the cell, we obtain a space $X \cup_\varphi (D^n \backslash 0)$ which is homotopy equivalent to X, because $X \subset X \cup_\varphi (D^n \backslash 0)$ is a strong deformation retract:

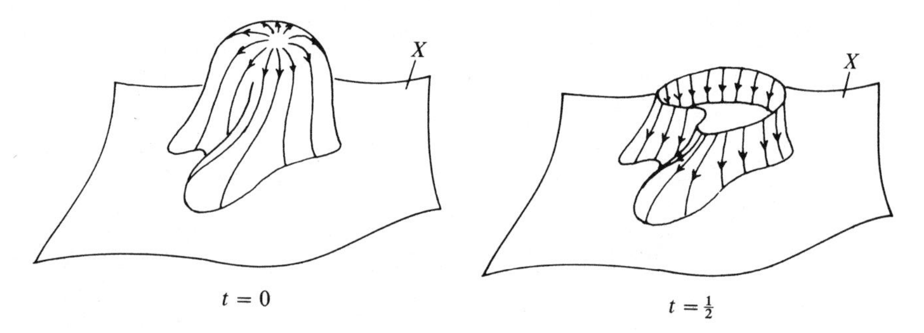

Example 3. Let $0 < k < n$. Look at $\mathbb{R}^{n+1}$ as $\mathbb{R}^k \times \mathbb{R}^{n-k+1}$ and consider in $S^n \subset \mathbb{R}^{n+1}$ the subspace $\sqrt{2}/2(S^{k-1} \times S^{n-k}) = \{(x, y) \mid \|x\|^2 = \|y\|^2 = \frac{1}{2}\}$, which is a "product of spheres". Then $\sqrt{2}/2(S^{k-1} \times S^{n-k})$ is a strong deformation retract of $S^n \backslash (S^{k-1} \times 0 \cup 0 \times S^{n-k})$.

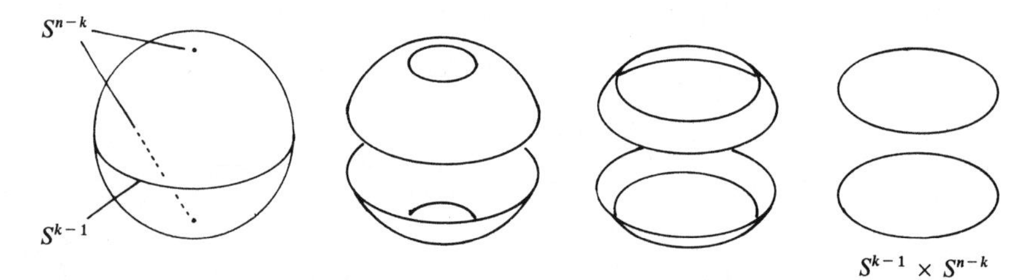

Example 4. A "figure eight" and a figure consisting of two circles connected by a line segment are homotopy equivalent, for they are both strong deformation retracts of the same space, a "thickened eight":

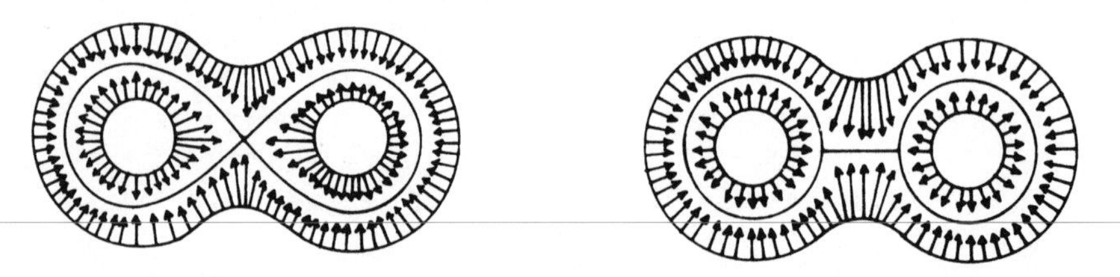

Example 5. Let M be a differentiable manifold with boundary. If the boundary ∂M is non-empty, it is obvious that $M \backslash \partial M$ cannot be a retract of M, since it is dense in M. But using an appropriate "collar neighborhood" one can see that M and $M \backslash \partial M$ have a common deformation retract. They are thus homotopy equivalent.

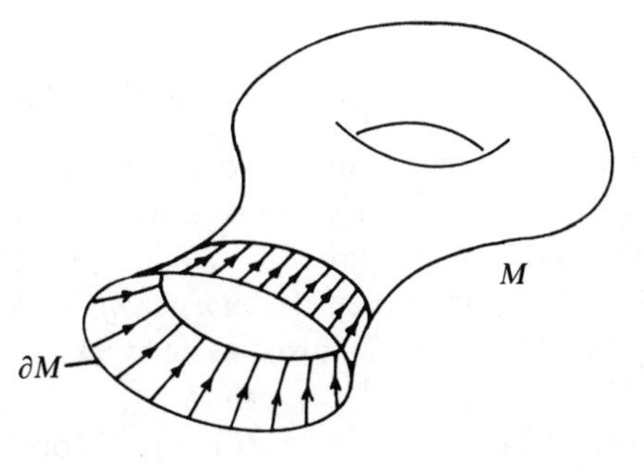

Example 6. Yet another example from differential topology, specifically from Morse theory. $M_a \cup_\varphi D^k$ is a strong deformation retract of M_b, in the notation of III, §7, Ex. 2 (cf. [14]).

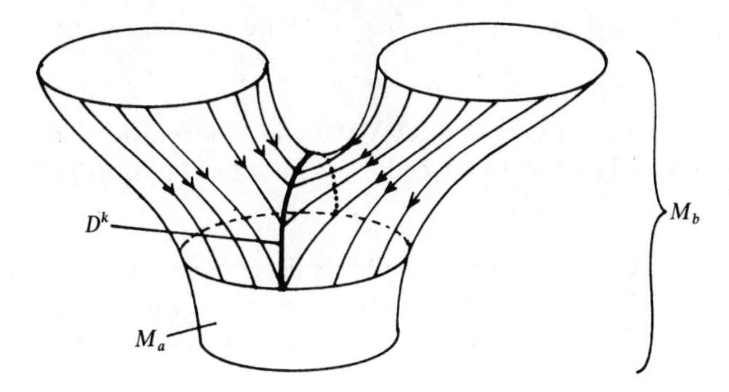

Example 7.

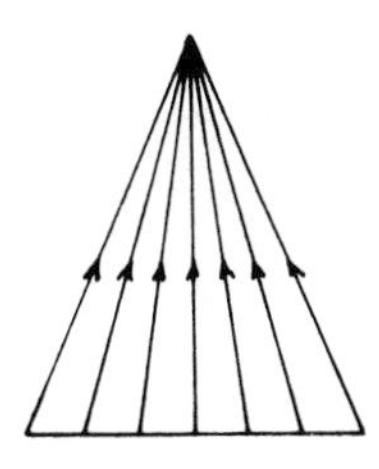

For every topological space X the cone CX is contractible: the vertex is a strong deformation retract of the cone.

§4. Categories

In order to be able to explain the essence and purpose of the concept of homotopy, I must say first what is understood by "algebraic topology", and this is best done by introducing an appropriate language, which also finds many other applications in mathematics: the language of categories and functors.

Definition (Categories). A *category* $\mathscr{C}$ consists of the following data:

(a) a class $\text{Ob}(\mathscr{C})$ of mathematical objects, called the *objects of the category*.
(b) a set $\text{Mor}(X, Y)$ for each pair (X, Y) of objects, where $\text{Mor}(X, Y)$ and $\text{Mor}(X', Y')$ are disjoint if the pairs (X, Y) and (X', Y') are distinct. The elements of $\text{Mor}(X, Y)$ are called the *morphisms of X into Y*. Notation: Instead of $f \in \text{Mor}(X, Y)$ we'll also write $f: X \to Y$, not necessarily implying that X, Y are sets and f is a map.
(c) a law of composition $\text{Mor}(X, Y) \times \text{Mor}(Y, Z) \to \text{Mor}(X, Z)$ for each triple of objects (X, Y, Z) (we write the law of composition as $(f, g) \mapsto g \circ f$, corresponding to the notation $X \xrightarrow{f} Y \xrightarrow{g} Z$ borrowed from sets and maps).

Data (a), (b) and (c) form a category if they fulfill the following axioms:

Axiom 1 (Associativity). If $X \xrightarrow{f} Y \xrightarrow{g} Z \xrightarrow{h} U$ are morphisms, then

$$h \circ (g \circ f) = (h \circ g) \circ f.$$

Axiom 2 (Identity). For every object X there is a morphism $1_X \in \text{Mor}(X, X)$ with the property $1_X \circ f = f$ and $g \circ 1_X = g$ for all morphisms $f: Y \to X$ and $g: X \to Z$.

Before proceeding any further I will first list some examples of categories. As long as the morphisms are maps and nothing is said about the law of composition, we'll always have in mind the usual composition of maps.

Example 1. The category $\mathscr{M}$ of sets:

(a) Objects: sets;
(b) Morphisms: maps.

Example 2. The topological category $\mathscr{T}\!op$:

(a) topological spaces;
(b) continuous maps.

By the way, Example 2′:

(a) topological spaces;
(b) arbitrary maps between topological spaces,

is also a category, but not a particularly interesting one.

Example 3. Category of groups:

(a) groups;
(b) group homomorphisms.

Example 4. Category of vector spaces over $\mathbb{K}$:

(a) $\mathbb{K}$-vector spaces;
(b) $\mathbb{K}$-linear maps.

Example 5. Category of topological vector spaces over $\mathbb{K}$:

(a) topological vector spaces over $\mathbb{K}$;
(b) continuous linear maps.

Example 6. The differential topological category $\mathscr{D}\!\mathit{ifftop}$:

(a) differentiable manifolds;
(b) differentiable maps.

Example 7. The category $\mathscr{V}\!\mathit{ect}\,(X)$ of vector bundles over a given topological space X:

(a) vector bundles over X;
(b) bundle homomorphisms, i.e. continuous, fiberwise linear maps over the identity in X:

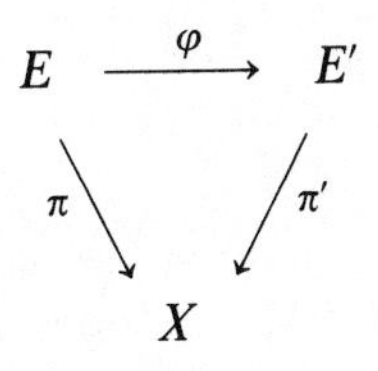

Example 8. The category of n-dimensional vector bundles over arbitrary topological spaces:

(a) n-dimensional vector bundles;
(b) "bundle maps", i.e. continuous, fiberwise isomorphic maps over continuous maps from one base space to another:

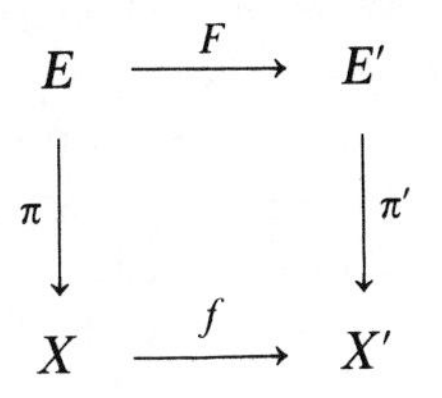

We could continue in this way for a long time, because practically every sort of mathematical structure admits structure-preserving or structure-compatible maps, and the category axioms don't require too much. There are examples by the dozen in algebra, analysis, topology as well as elsewhere.

The examples presented so far share the feature that their objects are sets with some additional structure, and the morphisms are maps with the usual composition (which is why we don't have to worry about associativity). But the concept of category goes further. As an illustration I'll give the following example, which may look somewhat strange:

Example 9. Let G be a group.

(a) only one object, which we call e;
(b) $\mathrm{Mor}(e, e) := G$;
(c) composition is the group multiplication.

A really important example of a category in which the morphisms are not maps is the following:

Example 10. The homotopy category $\mathscr{H}\mathit{top}$

(a) the objects are the topological spaces, as with $\mathscr{T}\mathit{op}$; but
(b) the morphisms are the homotopy classes of continuous maps;

$$\mathrm{Mor}(X, Y) := [X, Y]:$$

and
(c) the composition is defined via representatives of each class

$$[g] \circ [f] = [g \circ f].$$

*

After the definition and examples, a couple of complementary remarks. From the identity axiom it follows at once that for each object there is *exactly* one "identity" or "one" 1_X, because if $1'_X \in \mathrm{Mor}(X, X)$ has the same property, then $1'_X = 1'_X \circ 1_X = 1_X$. In the same way a morphism $f: X \to Y$ can have at most one *inverse* morphism $g: Y \to X$ (i.e. such that $f \circ g = 1_Y$ and $g \circ f = 1_X$), because if g' has the same property, $g' \circ (f \circ g) = g' \circ 1_Y$, hence from associativity $(g' \circ f) \circ g = 1_X \circ g = g = g' \circ 1_Y = g'$.

The morphisms that possess an inverse are called *isomorphisms* of the category, and two objects between which there is an isomorphism are called *isomorphic*. Thus in the topological category two spaces are isomorphic if they are homeomorphic, but in the homotopy category isomorphism means only that the spaces are homotopy equivalent.

To conclude, a last remark about a word used in the definition, which may already have momentarily puzzled you. There is good reason to speak only of the "class" $\mathrm{Ob}(\mathscr{C})$ of objects and not of the "set of objects of $\mathscr{C}$". You already know the contradictions to which the naive approach to set theory can lead, with its phrases like "the set of all sets". True, here we require that for a given category $\mathscr{C}$ the concept of objects be defined with enough precision (as in the case of topological spaces, for instance), but we don't expect all the objects of $\mathscr{C}$ that there ever have been, are and will be, to form a well-defined set that can be included in the usual set-theoretical operations. There are of course categories whose objects really do form a set; they are the so-called "small categories".

A formal and exact understanding of the meaning of "set" and "class" requires the use of axiomatic set theory. Here we satisfy ourselves with the warning that object classes are not to be taken as more than simply the concept of the objects in question.

§5. Functors

Definition (Covariant Functor). Let $\mathscr{C}$ and $\mathscr{D}$ be categories. By a covariant functor $\mathscr{F}: \mathscr{C} \to \mathscr{D}$ we mean a correspondence ("functor data") which associates to each object X of $\mathscr{C}$ an object $\mathscr{F}(X)$ of $\mathscr{D}$, and to each morphism

$$X \xrightarrow{\varphi} Y$$

of $\mathscr{C}$ a morphism

$$\mathscr{F}(X) \xrightarrow{\mathscr{F}(\varphi)} \mathscr{F}(Y)$$

of $\mathscr{D}$, such that the following "functor axioms" hold: $\mathscr{F}$ respects the category structure, i.e.

(1) $\mathscr{F}(1_X) = 1_{\mathscr{F}(X)}$; and
(2) $\mathscr{F}(\varphi \circ \psi) = \mathscr{F}(\varphi) \circ \mathscr{F}(\psi)$ for all etc. (it's clear what for).

Definition (Contravariant Functor). Analogously, but with the difference that $\mathscr{F}$ now reverses the direction of morphisms: To each $X \xrightarrow{\varphi} Y$ is associated a morphism

$$\mathscr{F}(X) \xleftarrow{\mathscr{F}(\varphi)} \mathscr{F}(Y)$$

(this stands for an element $\mathscr{F}(\varphi) \in \operatorname{Mor}(\mathscr{F}(Y), \mathscr{F}(X))$). The identity axiom remains the same for contravariant functors, but the composition axiom must be written now as $\mathscr{F}(\varphi \circ \psi) = \mathscr{F}(\psi) \circ \mathscr{F}(\varphi)$, because

$$X \xrightarrow{\psi} Y \xrightarrow{\varphi} Z$$

is taken to

$$\mathscr{F}(X) \xleftarrow{\mathscr{F}(\psi)} \mathscr{F}(Y) \xleftarrow{\mathscr{F}(\varphi)} \mathscr{F}(Z).$$

Remark. To be sure, the difference between the two concepts is only formal, because every category has a "dual category" with the same objects, defined by $\operatorname{Mor}^{\text{dual}}(X, Y) := \operatorname{Mor}(Y, X)$ and $\varphi \circ^{\text{dual}} \psi := \psi \circ \varphi$, and with this notation a contravariant functor from $\mathscr{C}$ into $\mathscr{D}$ is nothing but a covariant functor from $\mathscr{C}$ to $\mathscr{D}^{\text{dual}}$. But in view of the relevant examples it is more practical and natural to talk about contravariant functors than about dual categories.

For trivial examples take the identity functors $\mathrm{Id}_{\mathscr{C}}: \mathscr{C} \to \mathscr{C}$, which is evidently covariant, and the constant functors $\mathscr{C} \to \mathscr{D}$, which associate to every object the fixed object Y_0, and to every morphism the identity 1_{Y_0}. The constant functors can be seen as either co- or contravariant.

It would sometimes be inconvenient not to be allowed to call such correspondences functors, but they aren't exactly interesting. Somewhat more noteworthy are the "forgetful functors", for instance the covariant functor $\mathscr{T}\!op \to \mathscr{M}$ which associates to every topological space X the set X, and to every continuous map $f: X \to Y$ the map $f: X \to Y$. One often has reason to consider such functors from a category with more structure into one with less; all they do is "forget" the richer structure of the domain category.

The first examples of functors with real mathematical content are to be found in linear algebra. For instance, let $\mathbb{K}$ be a field and $\mathscr{V}$ the category of $\mathbb{K}$-vector spaces and linear maps. Then the concept of the "dual space" of a vector space gives us a contravariant functor $*: \mathscr{V} \to \mathscr{V}$ in a canonical way: To each object V is associated the dual space $V^* := \{\varphi: V \to \mathbb{K} \mid \varphi \text{ is linear}\}$, and to each linear map $f: V \to W$ the dual map $f^*: W^* \to V^*$, $\alpha \mapsto \alpha \circ f$. Then $\mathrm{Id}_V^* = \mathrm{Id}_{V^*}$ and $(f \circ g)^* = g^* \circ f^*$, and $*$ is a (contravariant) functor.

Functors which have not only content but mathematical power as well are admittedly harder to come by, but more about that later. This section is just to introduce the concept, and I will close it with a simple example that has to do with homotopy. Recall that $[X, Y]$ denotes the set of homotopy classes of continuous maps $X \to Y$. Then

Example. Let B be a topological space. Then $[.., B]$ defines in a canonical way a contravariant functor from the homotopy category into the category of sets and maps, as follows: To every topological space X we associate the set $[X, B]$ and to every morphism $[f] \in [X, Y]$ of the homotopy category we associate the map $[f, B]: [Y, B] \to [X, B]$ defined by $[\varphi] \mapsto [\varphi \circ f]$.

§6. What Is Algebraic Topology?

In a nutshell: Algebraic topology is solving topological problems using algebraic methods. — Now what on earth can that mean? Well, we're going to explain it in more detail.

What we call today algebraic topology was from the older, simpler and more obvious point of view the finding, calculation and application of *invariants*. A correspondence χ that associates to every X in a given class of geometrical objects a number $\chi(X)$, is called an invariant if $X \cong Y$ always implies $\chi(X) = \chi(Y)$. What sort of objects and isomorphisms "$\cong$" this definition refers to depends on the particular case under consideration; we speak for instance of "topological invariants" when $\cong$ means homeomorphism, of "diffeomorphism invariants" when $\cong$ means diffeomorphism, and so on.

No doubt the oldest non-trivial example of such an invariant is the Euler number or Euler characteristic of finite polyhedra. Let P be a polyhedron in $\mathbb{R}^n$, consisting of a_0 vertices, a_1 edges, a_2 two-dimensional "sides" and so on (we don't have to get into the exact definition here). Then

$$\chi(P) := \sum_{i=0}^{n} (-1)^i a_i$$

is called the *Euler number* of the polyhedron P, and the following non-trivial invariance theorem holds: *The Euler number is a topological invariant.* This gives us right away a topological invariant for all topological spaces X that are homeomorphic to finite polyhedra: $\chi(X)$ is well defined (by the invariance theorem) as the Euler number of a polyhedron homeomorphic to X.

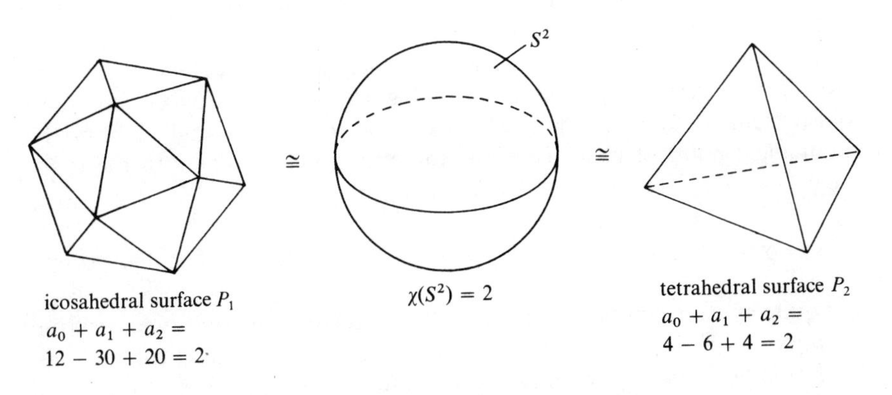

How can one apply such invariants to the solution of geometrical problems? Here is one example. Consider the following surfaces X and Y:

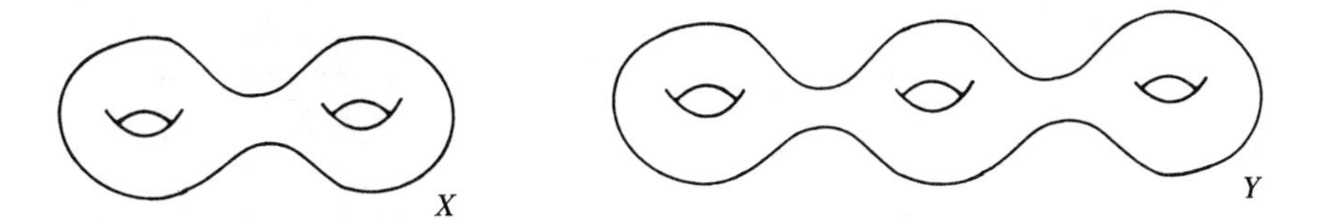

Are X and Y homeomorphic? They are both compact and connected, and of course Hausdorff as well: so they cannot be immediately distinguished, and just the fact that we aren't able to construct a homeomorphism between them doesn't prove anything. But $X \not\cong Y$ should somehow follow from $2 \neq 3$, shouldn't it? And such is indeed the case, because the computation of Euler characteristics gives $\chi(X) = -2$, but $\chi(Y) = -4$. Thus X and Y cannot be homeomorphic, qed.

Other topological invariants that were known around the turn of the century, before the introduction of the modern viewpoint, were for example the so-called "Betti numbers" b_i and the "torsion coefficients". The Betti numbers are connected with the Euler characteristic via $\chi = \sum(-1)^i b_i$, but two spaces can have the same Euler characteristic and different Betti numbers, so in this sense the Betti numbers are a "finer" invariant, and one can extract more information from them.

From the modern point of view algebraic topology is the finding, calculation and application of functors from "geometric" categories (e.g. $\mathcal{Top}$, $\mathcal{Difftop}$, ...) into "algebraic" categories (e.g. the category of groups, of rings, ...).

A fundamental example, that paved the way in the development of the modern point of view, is that of "homology": for every $k \geq 0$ one has the (covariant) k-dimensional homology functor H_k from the topological category into the category of abelian groups. Mathematicians gradually acquired greater skill in inventing appropriate functors, and there are now a great many functors in use in algebraic topology, some co- and some contravariant. In particular the somewhat vague concept of "geometric" category must now be understood in a much wider context. In analysis, for example, one investigates "geometric" objects and categories (complex spaces and manifolds, Riemann surfaces etc.), which can be (and are!) studied from the topological point of view, using the forgetful functor to pass to the topological category and functors defined thereon:

$$\text{Category of complex spaces} \xrightarrow[\text{functor}]{\text{forgetful}} \mathcal{Top} \xrightarrow{H_k,\text{ e.g.}} \text{Category of abelian groups.}$$

But quite apart from that, complex analysis directly constructs functors from "complex analytic" categories into algebraic ones, using analytical methods. Such analytically defined functors are often "finer" than the topological ones, since they do not "forget" the complex structure.

Well, but what is the good of all those functors? Now, first we want to remark that the functor axioms imply "invariance" in the following sense: If H is a functor and $f: X \to Y$ an isomorphism, then $H(f): H(X) \to H(Y)$ (resp. $H(Y) \to H(X)$, in the contravariant case), is an isomorphism as well, because if g is the inverse morphism to f, $H(g)$ is obviously inverse to $H(f)$. In particular $X \cong Y$ always implies $H(X) \cong H(Y)$: such "invariance theorems" come along free with every functor, and one can use the isomorphism classes of these algebraic objects to distinguish between geometric objects exactly as in the case of numerical invariants. In fact the classical invariants, too, can be obtained as invariants of these algebraic objects: for example, the i-th Betti number is the rank of the i-th homology group: $b_i(X) = \operatorname{rk} H_i(X)$, and $\chi(X) = \sum_{i=0}^{\infty}(-1)^i \operatorname{rk} H^i(X)$, etc. This just to show that the modern functors are no worse than the classical invariants. But the modern point of view of algebraic topology has in fact advantages over the older one, as we are now going to explain.

I'll not elaborate on the fact that the algebraic objects $H(X)$ contain in general more information than the invariants, that they make possible distinctions between geometrical objects for which the invariants would give us just a bleak "∞", or not even be defined—so I can go right to the center of things, and that is: The functors give not only information about the geometrical objects, but also about the geometrical morphisms, about the maps! An example will illustrate what this means.

A geometrical problem often boils down to whether a given continuous, surjective map $\pi: X \to Y$, about which we already know for instance that it is not injective (hence certainly does not possess an inverse) admits a *section*, i.e. a continuous map $\sigma: Y \to X$ such that $\pi \circ \sigma = \mathrm{Id}_Y$.

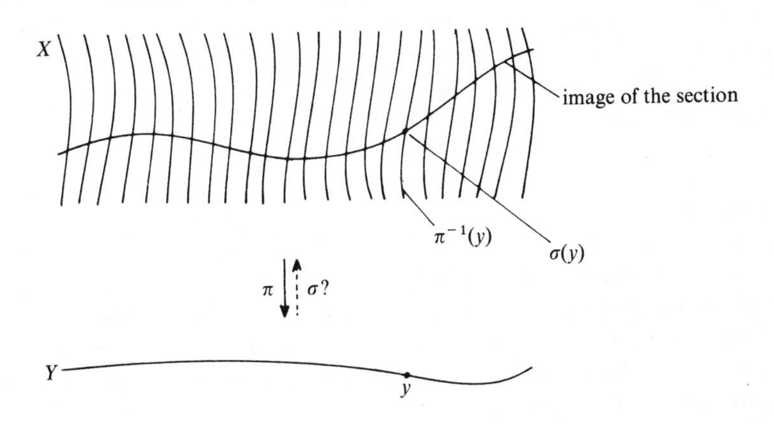

An analogous problem is often met in many other categories, the idea being, generally speaking, to decide whether a morphism $\pi: X \to Y$ has a "right inverse", i.e. a morphism $\sigma: Y \to X$ with $\pi \circ \sigma = 1_Y$. Let's say for instance we have a continuous surjective map π from S^3 onto S^2. Can π have a section? You can see that it's no good calculating invariants for S^2 and S^3, because it's not a matter of their being equal or different. But things look different if we apply an appropriate functor: If there is a σ such that the composition

$$S^2 \xrightarrow{\sigma} S^3 \xrightarrow{\pi} S^2$$

is the identity, then, by the functor axioms, the composition

$$H(S^2) \xrightarrow{H(\sigma)} H(S^3) \xrightarrow{H(\pi)} H(S^2)$$

must also be the identity. But using two-dimensional homology, for instance, we have $H_2(S^2) \cong \mathbb{Z}$ and $H_2(S^3) = 0$, and thus the composition

$$\mathbb{Z} \xrightarrow{H(\sigma)} 0 \xrightarrow{H(\pi)} \mathbb{Z}$$

must be the identity on $\mathbb{Z}$, which is obviously impossible. So no map from S^3 to S^2 can have a section.

That was just a simple application example, but typical of the superiority of the functorial point of view in all questions which have to do with maps. But even if one were interested only in the geometrical objects themselves, the old point of view in algebraic topology wouldn't have developed much further, because the study of spaces and the study of maps depend so much on one another that any development unilaterally concentrated on spaces would have led to a dead end.

§7. Homotopy—What For?

After all those preparations I can now give a reasonable answer to this question, and I will in fact adduce two interconnected main reasons for the usefulness of the notion of homotopy. The first is the homotopy invariance of most algebraic topological functors. A functor defined on the category of topological spaces is called homotopy invariant if $f \simeq g$ always implies $H(f) \cong H(g)$. Such functors can be thought of as being actually defined on the homotopy category, their application to $\mathcal{Top}$ arising by composition with the canonical functor: $\mathcal{Top} \to \mathcal{Htop} \to \mathcal{A}$. From the functor axioms it follows of course that a homotopy invariant functor assigns isomorphic objects to homotopy equivalent spaces: $X \simeq Y$ implies $H(X) \cong H(Y)$. Categories other than that of topological spaces can also, with a suitably modified notion of homotopy, admit the concept of homotopy invariant functors, and, as I said, many (though not all) functors in algebraic topology have this property.

This is not implausible anyway, since homotopy invariance means that for every homotopy h the morphisms $H(h_t)$ do not depend on t, which, due to the connectedness of the interval $[0, 1]$, means simply that $H(h_t)$ is *locally* constant relative to t. In this form: "For sufficiently small deformations of a map the associated algebraic morphism is not changed", homotopy invariance lies in the very nature of correspondences which coarsen continuous behavior into algebraic features.

All right, but why is this so important? Well, *because the computability of functors is based to a large extent on this fact*! The best of functors is no good if it cannot be calculated. Applying the definition directly would be too complicated, but the task can often be simplified by using homotopy invariance to pass to homotopically equivalent spaces.

In fact, explicit calculations are carried out directly from the definition only for a couple of extremely simple standard spaces (e.g. the one-point space, S^1 and the like), and after that one applies "laws", among which the homotopy invariance is one of the most important (others are for instance the Mayer–Vietories principle, long exact sequences, spectral sequences . . .).

The second main reason for the usefulness of the notion of homotopy is the possibility of "reducing" some geometric problems to homotopy problems. When we apply an algebraic topological functor to a geometric situation, i.e. to all spaces and maps occurring in it, we get in general a strongly simplified, but for that very reason more transparent, algebraic "representation" of the geometric situation. Geometric questions are thus translated into (simpler) algebraic ones, and by answering the latter we may get at least some partial information about the former, for instance: For $f: X \to Y$ to have a right inverse it is certainly necessary that $H(f): H(X) \to H(Y)$ has, but it doesn't work the other way around—this condition is in general not sufficient. The functor does not reflect all the essential features of the geometric problem, but only one aspect of it. Now the homotopy category stands half-way, so to speak, between the extremes of topological intractability and algebraic oversimplification. On the one hand it is quite "fine" and is close to the topological category, as attested by the homotopy invariance of so many functors. Thus homotopy conditions are sometimes really sufficient, and the original topological problem can be solved by solving its representation in the homotopy category. On the other hand, the latter is still coarse and algebraic enough not to make calculations entirely inaccessible. Loosely speaking, there are much fewer homotopy classes than maps, so that one can to a certain extent get a general view of $[X, Y]$. For instance: There are many and very complicated closed curves $S^1 \to \mathbb{C}\backslash 0$, but $[S^1, \mathbb{C}\backslash 0] = \mathbb{Z}$ ("winding number").

An important ingredient for the algebraic manageability of the homotopy category are the so-called "homotopy groups" of a topological space, and I will digress from my main topic to include here their definition. To this effect we need a bit of notation, namely: A space with basepoint simply means a pair (X, x_0) consisting of a topological space X and a point $x \in X_0$. It is then clear what basepoint-preserving continuous maps $(X, x_0) \to (Y, y_0)$ are, ditto for basepoint-preserving homotopies between such maps. Denote by $[(X, x_0), (Y, y_0)]$ the set of such homotopy classes. If now N is a fixed point on the n-sphere S^n, $n \geq 1$, the north pole for instance, then

$$\pi_n(X, x_0) := [(S^n, N), (X, x_0)]$$

has a canonical group structure (abelian for $n \geq 2$) which I will presently explain, and is called the n-th homotopy group of (X, x_0).

The group law is most easily described by considering the n-sphere as the quotient $I^n/\partial I^n$, obtained from the cube $I^n := [0, 1]^n$ by collapsing the boundary $\partial I^n := \{(x_1, \ldots, x_n) \in I^n \mid \text{at least one } x_i \text{ is 0 or 1}\}$ (cf. p. 42 and p. 94). So choose once and for all a homeomorphism $I^n/\partial I^n = S^n$ taking the point ∂I^n to the north pole. Then the continuous maps $(S^n, N) \to (X, x_0)$ are exactly the continuous maps $I^n \to X$ that take the whole boundary ∂I^n to x_0. Now if α, β are two such maps, define a map $[0, 2] \times [0, 1]^{n-1} \to X$ in the obvious way, given by α on the left half and by β on the right half, and compose this map with another map $I^n \to [0, 2] \times I^{n-1}$ stretching the first coordinate by a factor of two:

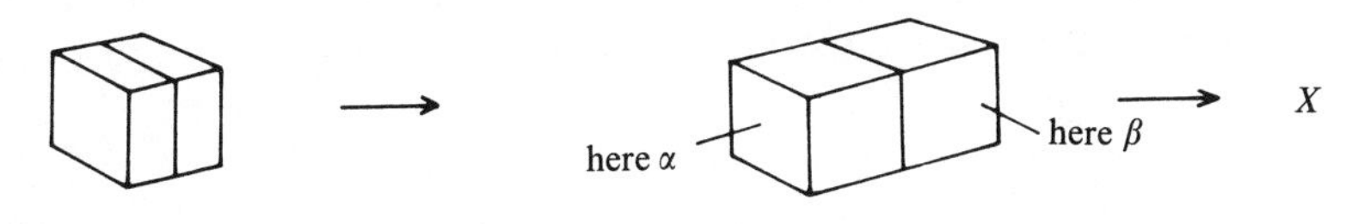

The composition then represents another element of $\pi_n(X, x_0)$ which is defined to be the composition $[\alpha][\beta]$ of the elements $[\alpha]$, $[\beta] \in \pi_n(X, x_0)$. (Cf., for example, [11], p. 5).

*

Picking up the thread again, this midway position of homotopy relative to topology and algebra made it possible to take important geometric problems and first reformulate them as homotopy problems and then solve them, completely or in part, using homotopy-theoretical calculus. Those things are admittedly way above what can be done with the tools and on the level of this book, and a critical observer may find it outrageous to talk about them here. But this shall not stop me from mentioning at least *one* example to satisfy a bit your by now whetted curiosity.

Two n-dimensional, compact differentiable manifolds without boundary M_1 and M_2 are called "bordant" if there is an $(n + 1)$-dimensional compact manifold *with* boundary W whose boundary is the disjoint union of M_1 and M_2:

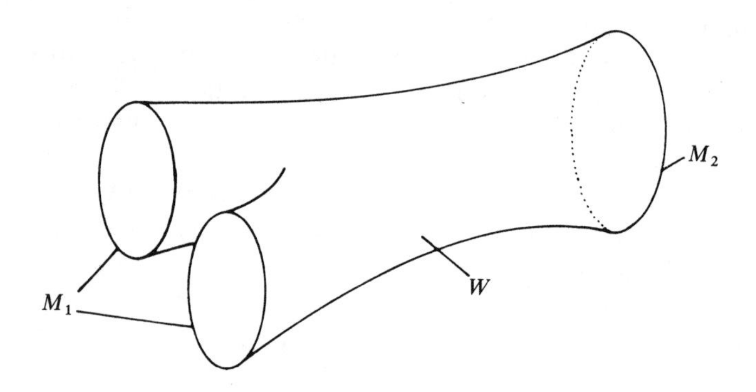

"Bordant" is an equivalence relation, and the equivalence classes, called "bordism classes", form an abelian group $\mathfrak{N}_n$ with the composition law defined by disjoint union. Problem: determine those groups.

Solving this problem would be equivalent to classifying n-dimensional manifolds up to bordism. Now bordism is, in comparison to diffeomorphism, a quite coarse relation, but in differential topology there are lots of reasons to welcome coarser classifications as well, because to this day little enough is known about the classification of higher dimensional manifolds up to diffeomorphism, and around the time of our story, the early fifties, nothing whatever was known. Besides bordism is not so coarse a relation as it might seem at first sight; it preserves some important properties of manifolds, and in any case the classification up to bordism subsequently proved very fruitful and useful.

Now this problem does not look very accessible, given the above-mentioned circumstances! René Thom solved it in 1954 using homotopy-theoretical methods. I will not include here the result (see [18]), but the $\mathfrak{N}_n$ are certain finite abelian groups, quite different for different n, and not at all to be guessed using only geometric intuition. But I shall say a few words about the homotopic formulation of the problem. In Chapter III, §4 we talked about the Grassmann manifold $O(n+k)/O(k) \times O(n)$, whose points are the k-dimensional vector subspaces of R^{n+k}. Over this manifold there is a canonical k-dimensional vector bundle ("Grassmann bundle") whose fiber over a "point" $\xi \subset \mathbb{R}^{n+k}$ is exactly the k-dimensional space ξ itself (or rather $\xi \times \{\xi\}$, since the fibers must be mutually disjoint). Denote by $M_n O(k)$ the *Thom space* of this bundle (see III, §6 example 5). Thom was able to prove that *for k big the homotopy group $\pi_{n+k}(M_n O(k))$ is isomorphic* to $\mathfrak{N}_n$. Calculating these homotopy groups is then the homotopy problem to which the bordism problem could be reduced; and the homotopy problem could be solved by Thom using the new methods with which J. P. Serre had not long before achieved a great breakthrough in homotopy theory, which had known a long period of stagnation before that.

I would greatly like to outline here the "Pontrjagin–Thom construction" which Thom used to perform the transformation of the bordism problem into a homotopy problem, and to tell how this construction is connected with an earlier one by Pontrjagin (1938) and a still earlier one by Hopf (1926)—a very interesting and instructive detail of the development of modern topology. But for once I will resist the temptation....

CHAPTER VI

The Two Countability Axioms

§1. First and Second Countability Axioms

This short chapter is linked directly to the basics. We recall that a set $\mathfrak{B}$ of open sets in X is called a basis of the topology of X if *every* open set is the union of sets in $\mathfrak{B}$. Now we add to this notion that of "neighborhood basis":

Definition (Neighborhood Basis). Let X be a topological space, $x_0 \in X$. A set $\mathfrak{U}$ of neighborhoods of x_0 is called a neighborhood basis of x_0 if every neighborhood of x_0 contains a neighborhood in $\mathfrak{U}$.

Example. The set of all neighborhoods of x is of course an (uninteresting) neighborhood basis. But now let $X = \mathbb{R}^n$. The set of balls $K_{1/n}(x_0)$ with radius $1/n, n = 1, 2, ..$ around x_0 forms a (countable!) neighborhood basis of x_0.

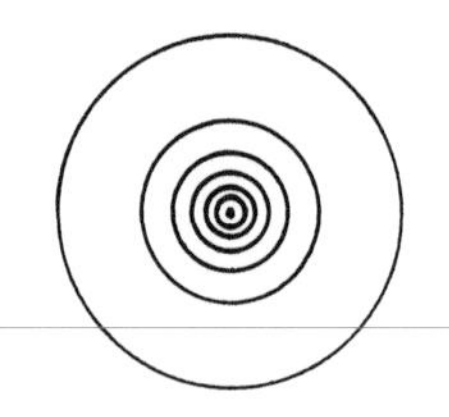

Definition (Countability Axioms). A topological space satisfies the *first countability axiom*, and is called *first countable*, if every point possesses a countable neighborhood basis. It satisfies the *second countability axiom*, and is called *second countable*, if it possesses a countable basis for the topology.

Obviously the second axiom is the stronger one; the sets in a countable basis that contain x_0 form of course a countable neighborhood basis for x_0. Both axioms have the property of being inherited by subsets. $\mathbb{R}^n$ and thus all its subspaces satisfy both axioms (all balls with rational radius and rational center coordinates form a countable basis for the topology). Metrizable spaces satisfy at least the first axiom: If d is a metric, the d-balls $K_{1/n}(x_0)$ form a countable neighborhood basis of x_0.

To get a better look at the difference between the two axioms, we'll consider some examples of first countable spaces that are not second countable. Uncountable discrete spaces, which are trivially in this category, are, of course, uninteresting in themselves, but in looking for better examples it is useful to

Note. *If a topological space has an uncountable discrete subset, it cannot be second countable.*

Example 1. Let $C(\mathbb{R})$ be the Banach space of bounded continuous functions on $\mathbb{R}$ with the supremum norm. Then $C(\mathbb{R})$ is first countable (being a metric space), but not second countable.

Proof. Define for every real number x expressed as a decimal a continuous bounded function f_x whose value for $n \in \mathbb{Z}$ is the n-th decimal after the decimal point.

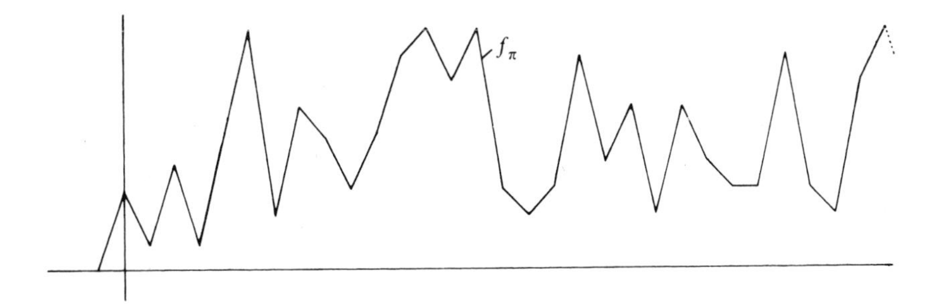

Then $\|f_x - f_y\|$ is always ≥ 1 for $x \neq y$, which means $\{f_x | x \in \mathbb{R}\}$ is an uncountable discrete subspace of $C(\mathbb{R})$, and $C(\mathbb{R})$ cannot be second countable.

□

Example 2. Let H be an "inseparable" Hilbert space, i.e. one for which there is no countable Hilbert basis. A Hilbert basis $\{e_\lambda\}_{\lambda \in \Lambda}$ will then have an uncountable index set, and since $\|e_\lambda - e_\mu\| = \sqrt{2}$ for $\lambda \neq \mu$, it follows as above that H does not satisfy the second axiom, but since it is a metric space it satisfies the first.

§2. Infinite Products

Of course we want to see also a topological space that satisfies neither countability axiom, and I use the question as an opportunity to talk for the first time about products of arbitrarily many topological spaces, a topic we'll discuss again in Chapter X.

By the *product* $\prod_{\lambda \in \Lambda} X_\lambda$ of a family $\{X_\lambda\}_{\lambda \in \Lambda}$ of sets we mean the set of families $\{x_\lambda\}_{\lambda \in \Lambda}$ of elements with $x_\lambda \in X_\lambda$ for all $\lambda \in \Lambda$, that is

$$\prod_{\lambda \in \Lambda} X_\lambda := \{\{x_\lambda\}_{\lambda \in \Lambda} | x_\lambda \in X_\lambda\}.$$

For $\mu \in \Lambda$ a fixed index, the projection $\pi_\mu : \prod_{\lambda \in \Lambda} X_\lambda \to X_\mu$ on the μ-th factor is defined by $\{x_\lambda\}_{\lambda \in \Lambda} \mapsto x_\mu$, and x_μ is also called the μ-th component of the point $\{x_\lambda\}_{\lambda \in \Lambda} \in \prod_{\lambda \in \Lambda} X_\lambda$. For $\lambda = \{1, \ldots, n\}$ it is of course better to write $(x_1, \ldots, x_n)$ instead of $\{x_\lambda\}_{\lambda \in \{1, \ldots, n\}}$, and then the above notations correspond to the familiar ones for finite cartesian products $X_1 \times \cdots \times X_n$.

Definition (Product Topology). Let $\{X_\lambda\}_{\lambda \in \Lambda}$ be a family of topological spaces. The *product topology* is defined as the coarsest topology relative to which the projections on the individual factors are all continuous. Together with this topology the set $\prod_{\lambda \in \Lambda} X_\lambda$ is called the *product space*.

The inverse images of open sets under the projections will be called "open cylinders".

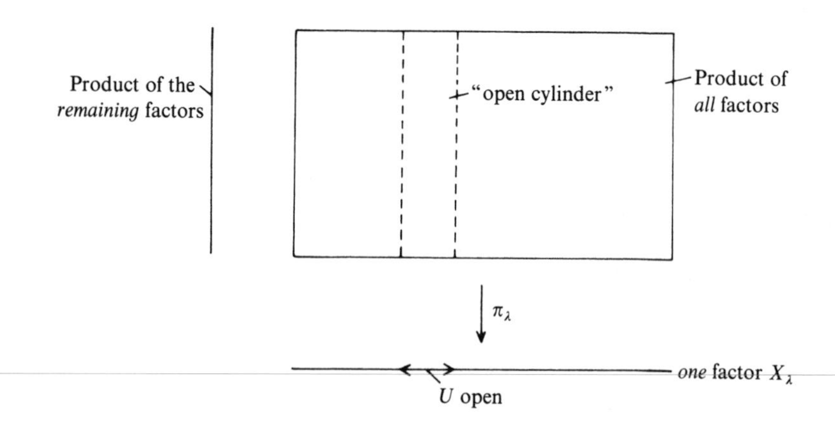

and the intersections of finitely many open cylinders will be called open boxes. Thus the open cylinders form a subbasis $\{\pi_\lambda^{-1}(U) | \lambda \in \Lambda, U \subset X_\lambda \text{ open}\}$ for the product topology, and the open boxes form a basis

$$\{\pi_{\lambda_1}^{-1}(U_1) \cap \cdots \cap \pi_{\lambda_r}^{-1}(U_r) | \lambda_1, \ldots, \lambda_r \in \Lambda, U_{\lambda_i} \subset X_{\lambda_i} \text{ open}\}.$$

One can also say that a subset of the product is open in the product topology if every point in the set is contained in an open box contained in the set.

If all the factors are the same, say $X_\lambda = X$ for all $\lambda \in \Lambda$, we also write X^Λ instead of $\prod_{\lambda \in \Lambda} X$. The elements of X^Λ are then simply (arbitrary) maps $\Lambda \to X$.

Returning now to the countability axioms:

Example 1. If Λ is uncountable and every X_λ is non-trivial (meaning simply that it has some open set other than $\varnothing$ and X_λ), the product $\prod_{\lambda \in \Lambda} X_\lambda$ is not first, and hence not second countable.

PROOF. For every λ choose an open set U_λ in X_λ that is neither $\varnothing$ and X_λ, and choose $x_\lambda \in U_\lambda$. If the point $\{x_\lambda\}_{\lambda \in \Lambda}$ had a countable neighborhood basis, it would have one consisting of open boxes. But only countably many λ can be "involved" in countably many open boxes. Take a λ that is not involved in any of the boxes. Then none of the boxes of the neighborhood basis will fit in $\pi_\lambda^{-1}(U_\lambda)$, a contradiction. qed. □

Example 2. An ∞-dimensional Hilbert space with the weak topology (the coarsest one for which the linear functionals, i.e. here the maps

$$\langle v, .. \rangle : H \to \mathbb{K}, v \in H,$$

are still continuous) is not first countable.

The proof is similar to that of Example 1, even in the separable case, for although H then has a countable Hilbert-basis, it has no countable vector space basis. Cf. [15], p. 379.

§3. The Role of the Countability Axioms

The first countability axiom has to do with *convergence of sequences*.

Notation. Instead of saying "there is an n_0 such that $x_n \in U$ for all $n \geq n_0$", we'll say "the sequence $(x_n)_{n \geq 1}$ eventually stays in U", partly because it is shorter, but also because it is more suggestive.

If $f: X \to Y$ is continuous and $\lim x_n = a$ in X, we have $\lim f(x_n) = f(a)$ in Y. This is a well-known fact and entirely trivial, for if U is a neighborhood of $f(a)$, then $f^{-1}(U)$ is a neighborhood of a, hence the sequence $f(x_n)$

eventually stays in $f(U)$ and (x_n) stays in U. Moreover, if X is a subspace of $\mathbb{R}^n$, the converse also holds: $f: X \to Y$ is continuous *if and only if* every convergent sequence is taken into a sequence that converges to the limit of the first. This characterization of continuity (sequential continuity, one might say) does not hold, however, for other spaces: the convergence of image sequences to the right point is not sufficient in general to guarantee continuity, as will be shown by the following

Example. Let X be the set of continuous functions $[0, 1] \to [-1, 1]$ endowed with the product topology, i.e. the topology inherited from

$$X \subset [-1, 1]^{[0,1]} = \prod_{\lambda \in [0,1]} [-1, 1].$$

As a set X is then the same as the unit ball in the Banach space $C[0, 1]$, but the topology under consideration is entirely different. What does it mean for a series to converge in X? In general, what does convergence mean in a product space? A sequence in $\prod_{\lambda \in \Lambda} X_\lambda$ converges to a if and only if it eventually stays in every open box around a, or again if and only if it remains in every open cylinder around a, which is equivalent to saying that it converges componentwise to a. Convergence in our example function space X is then nothing more than the familiar pointwise convergence: $\lim \varphi_n = \varphi$ means $\lim \varphi_n(\lambda) = \varphi(\lambda)$ for all $\lambda \in [0, 1]$.

Every continuous function on the interval $[0, 1]$ is *a fortiori* square-integrable, and thus we have a canonical map $X \to L^2[0, 1]$, $\varphi \mapsto \varphi$ from X into the Hilbert space of square-integrable functions on $[0, 1]$. Now this map is sequentially continuous (using, for instance, the Lebesgue convergence theorem), but it is not continuous. If it were, there would be for every $\varepsilon > 0$ an open box K around the origin in $[-1, 1]^{[0,1]}$ such that $\int_0^1 \varphi^2\, dx < \varepsilon$ for all $\varphi \in K \cap X$; but "belonging to K" is a condition about the value of φ on certain finitely many points in $[0, 1]$, and such a condition cannot prevent $\int_0^1 \varphi^2\, dx$ from being arbitrarily close to 1:

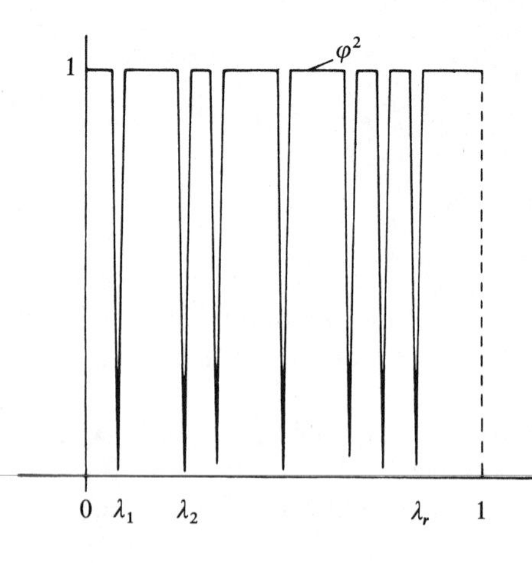

Proposition 1. *If X is first countable and Y is an arbitrary topological space, a map $f: X \to Y$ is continuous if and only if it is sequentially continuous.*

PROOF. Let f be sequentially continuous, $a \in X$ and U a neighborhood of $f(a)$. We must show: There is a neighborhood V of a with $f(V) \subset U$. Assume no V satisfies this condition, in particular the finite intersection $V_1 \cap \cdots \cap V_n$ of V_i in a countable neighborhood basis of a. Choose $x_n \in V_1 \cap \cdots \cap V_n$ with $f(x_n) \notin U$. Then $(x_n)_{n \geq 1}$ converges towards a, because there is a V_i contained in every neighborhood of a, and $x_n \in V_i$ for all $n \geq i$. But of course the image sequence does not converge towards $f(a)$, because it does not penetrate U at all, contradicting the sequential continuity of f. qed. □

More important than sequential continuity is perhaps the concept of sequential compactness, and here too the first countability axiom plays a decisive role.

Definition (Sequentially Compact). A topological space X is called sequentially compact if every sequence in X has a convergent subsequence.

Very often it would be desirable that compactness and sequential compactness were the same, be it that one needs convergent subsequences, or the other way around: one knows more about sequences than about open covers, a case which occurs particularly often in function spaces. The concepts, however, are not the same, and it is not even true in general that either of them implies the other. Instead of giving examples for this, I will provide a general reference, as follows: Let $\mathscr{A}$ and $\mathscr{B}$ be two topological properties about which you want to know if "$\mathscr{A} \Rightarrow \mathscr{B}$" holds; and suppose you feel it would be too difficult or too unreliable or just plain too boring to verify it yourself. Then of course you take the first available topology book, look in the index for properties $\mathscr{A}$ and $\mathscr{B}$, and if $\mathscr{A} \Rightarrow \mathscr{B}$ does indeed hold, it is quite

GENERAL REFERENCE CHART

No. of example in text	T_0	T_1	T_2	$T_{2\frac{1}{2}}$	T_3	$T_{3\frac{1}{2}}$	T_4	T_5	Urysohn	Semiregular	Regular	Completely Regular	Normal	Completely Normal	Perfectly Normal	Compact	σ-Compact	Lindelöf	Countably Compact	Sequentially Compact	Weak. Count. Compact	Pseudocompact	Locally Compact	Strong Loc. Compact	σ-Locally Compact	Separable	Second Countable	First Countable	Count. Chain Cond.	Paracompact
1	1	1	1	1	1	1	1	1	1	1	1	1	1	1	1	1	1	1	1	1	1	1	1	1	1	1	1	1	1	1
2	1	1	1	1	1	1	1	1	1	1	1	1	1	1	1	0	1	1	0	0	0	0	1	1	1	1	1	1	1	1
3	1	1	1	1	1	1	1	1	1	1	1	1	1	1	1	0	0	0	0	0	0	0	1	1	0	0	0	1	0	1
4	0	0	0	0	1	1	1	1	0	0	0	0	0	0	0	1	1	1	1	1	1	1	1	1	1	1	1	1	1	1
5																														
6	0	0	0	0	1	1	1	1	0	0	0	0	0	0	0	0	1	1	0	0	1	0	1	1	1	1	1	1	1	1
7																														
8	1	0	0	0	0	0	0	0	0	0	0	0	0	0	0	1	1	1	1	1	1	1	1	1	1	1	1	1	1	1
9	1	0	0	0	0	0	0	0	0	0	0	0	0	0	0	0	1	1	0	0	0	1	1	0	1	1	1	1	1	0

likely that you will find this fact stated as a lemma. But if $\mathscr{A} \Rightarrow \mathscr{B}$ *does not* hold, your chances are slimmer, at least in general; but there is one book which is excellent exactly for such cases, namely, L. A. Steen and J. A. Seebach, *Counterexamples in Topology* [17]. In this book are individually described 143 of mostly very strange topological spaces, and at the end there is a "Reference Chart", a big table where you can see at a glance whether or not each example has each one of 61 (!) topological properties.

Now all you have to do is inspect the columns for $\mathscr{A}$ and $\mathscr{B}$, and you'll find in particular (returning to our original theme) examples of compact but not sequentially compact spaces and vice-versa. However, the following holds:

Proposition 2. *A first countable compact space is also sequentially compact.*

PROOF. Let $(x_n)_{n \geq 1}$ be a sequence in X. Using first only the compactness of X we notice that there must be a point $a \in X$ such that the sequence penetrates infinitely often into every neighborhood of a, otherwise every point x would have a neighborhood U_x which intersects the sequence only a finite number of times, and since $X = U_{x_1} \cup \cdots \cup U_{x_n}$ the sequence would eventually have no place to step next. Now if a has a countable neighborhood basis $\{V_i\}_{i \geq 1}$, we can obviously choose a subsequence $\{x_{n_k}\}_{k \geq 1}$ with $x_{n_k} \in V_1 \cap \cdots \cap V_k$, and this subsequence converges towards a, qed. □

Proposition 3. *For metric spaces the concepts "compact" and "sequentially compact" are actually synonymous.*

PROOF. Let X be a sequentially compact metric space and $\{U_\lambda\}_{\lambda \in \Lambda}$ an open cover that has no finite subcover. We want to derive a contradiction. For every $x \in X$ we choose a $\lambda(x)$ such that x is not only contained in $U_{\lambda(x)}$, but is actually deeply imbedded in it, in the sense that the radius r of the biggest open ball around x which is still contained in $U_{\lambda(r)}$ must be either greater than 1 or so big that the ball of radius $2r$ around x is not contained in any of the sets of the covering. It is obviously possible to choose such $\lambda(x)$.

Now we choose a sequence $(x_n)_{n \geq 1}$ inductively with

$$x_{n+1} \notin U_{\lambda(x_1)} \cup \cdots \cup U_{\lambda(x_n)}.$$

Observe that now the distance of a member x_i of the sequence to any of its successors is either greater than one or great enough that the ball around x_i with radius double the distance does not fit in any of the sets of the covering. Now let a be the limit of a subsequence and $1 > r > 0$, so that $K_r(a) \subset U_{\lambda(a)}$. Then the subsequence would have to stay eventually in $K_{r/5}(a)$, but then the points in the sequence would have to be crammed together closer than allowed by our construction, a contradiction. qed. □

*

So much for the first countability axiom. Where does one come across the second? In a very prominent place, the definition of "manifolds": an n-dimensional topological manifold is a *second countable* Hausdorff space locally homeomorphic to $\mathbb{R}^n$. In lots of mathematical fields the objects of study are topological manifolds with additional structure, as for instance in differential topology, Riemannian geometry, Lie groups, Riemann surfaces, etc., and in yet other fields one studies structures that are like manifolds in some ways, e.g. complex spaces, and which are also required to be second countable ([10], p. 18). So one can say that this is one of the fundamental axioms of the greater part of modern geometry and topology.

From the mere definition of manifold it is not immediate what we require this axiom for. But its technical significance will soon become clear. In fact, it makes it possible to find for every open cover $\{U_\lambda\}_{\lambda \in \Lambda}$ (in particular for every family of open neighborhoods $\{U_x\}_{x \in X}$) a countable subcover, and this is needed for the many inductive constructions and proofs where one starts with local knowledge (locally homeomorphic to $\mathbb{R}^n$!) and proceeds from $U_{x_1} \cup \cdots \cup U_{x_n}$ to $U_{x_1} \cup \cdots \cup U_{x_n} \cup U_{x_{n+1}}$. But the second countability axiom is not merely a technical convenience: if it were to be discarded, theorems of differential topology like for instance the metrizability of manifolds, the Whitney embedding theorem, the Sard theorem etc. would not hold anymore.

Now of course this in itself would not be sufficient reason to leave aside spaces that do not satisfy the axiom, but otherwise look like a manifold. Maybe they are actually quite interesting? .... But this does not seem to be the case, and at any rate there are no positive reasons why one should study such "manifolds".

To conclude the chapter I will mention a sort of "third countability axiom" that sometimes comes up, namely, separability.

Definition. A topological space is called separable if it contains a countable dense set.

This property is of a rather different nature from the first and second countability axioms in that it is not inherited by subsets. For an example, consider $\mathbb{R}^2$ with the topology generated by the quadrants $(x, y) + \mathbb{R}^2_{++}$; this is a separable space, since the set $\{(n, n) \mid n \in \mathbb{N}\}$, for instance, is dense in it. But on the other hand the "antidiagonal" $x + y = 0$ is an uncountable discrete subspace, hence non-separable.

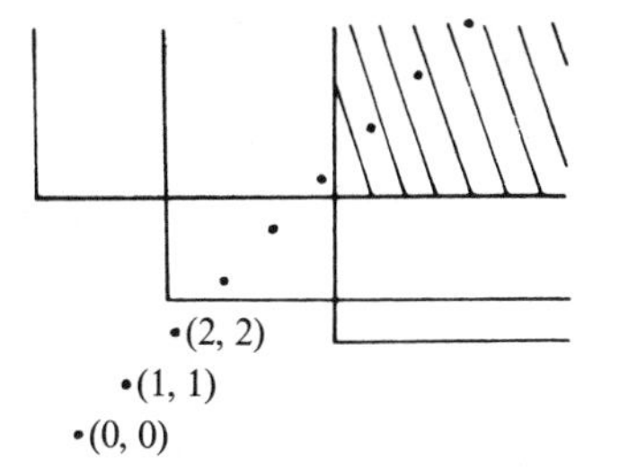

open

Now you will say, come now, this is a very pathological example. Granted! But in "reasonable" spaces, for instance metric spaces, the concept is dispensable, because metric spaces are separable if and only if they are second countable. In any case second countability implies separability, and for Hilbert spaces the definition coincides with the concept we've been already using: existence of a countable Hilbert basis.

CHAPTER VII
CW-Complexes

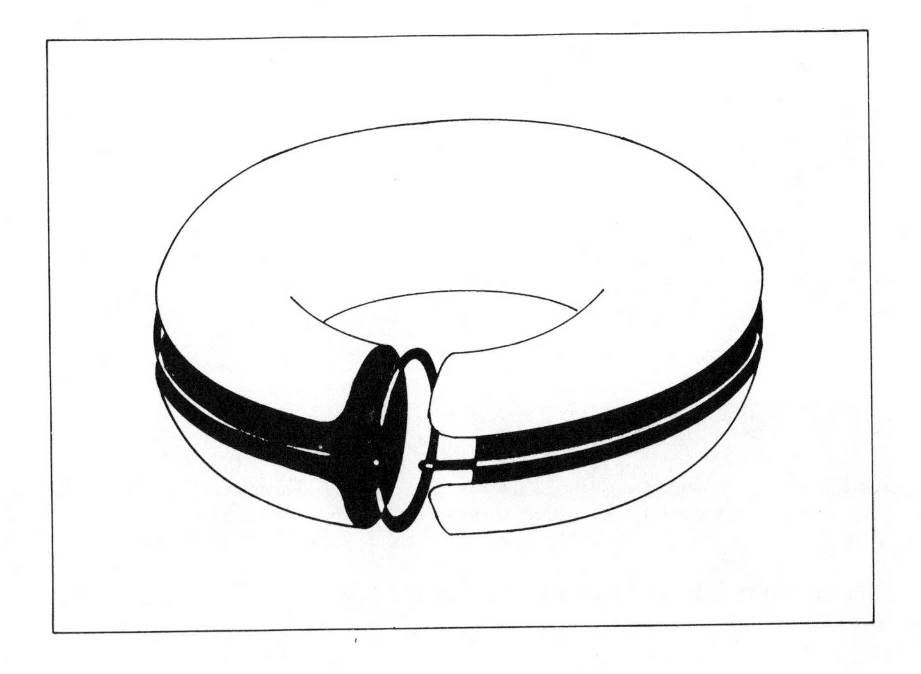

§1. Simplicial Complexes

Before we come to CW-complexes proper. I'd like to say something about their forerunners, the *simplicial* complexes. The language of point-set topology allows us to formulate concisely and consistently numerous problems which at first glance seem very diverse, and to submit them to a common intuitive presentation. But to the subsequent *solution* of such problems point-set topology, strictly speaking, has little to contribute. By far most solution methods come from algebraic topology. This was already known very early, and from the very beginning (i.e. approximately the turn of the century) it was a main endeavor of topology to develop the algebraic-topological machinery. Classical textbooks like Seifert–Threlfall, *Lehrbuch der Topologie* (1934) and Alexandroff–Hopf, *Topologie I* (1935) contain mainly algebraic topology, and the division into "point-set" and "algebraic" topology came about only after World War II, due to the accumulation of material.

Algebraic topology can well be said to start with simplices:

Definition (Simplex). By a *k-dimensional simplex* or *k-simplex* in $\mathbb{R}^n$ we mean the convex hull $s(v_0, \ldots, v_k)$ of $k + 1$ points in general position.

The convex hull of $v_0, \ldots, v_k$ is of course the set

$$\left\{\sum_{i=0}^{k} \lambda_i v_i \,|\, \lambda_0 + \cdots + \lambda_k = 1, \lambda_i \geq 0\right\},$$

and "general position" means that $(v_1 - v_0, \ldots, v_k - v_0)$ are linearly independent.

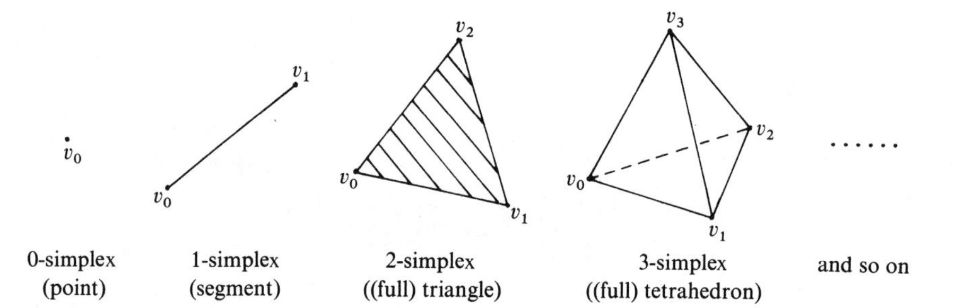

Nomenclature. The convex hull of a subset of $\{v_0, \ldots, v_k\}$ is called a subsimplex or a face of $s(v_0, \ldots, v_k)$:

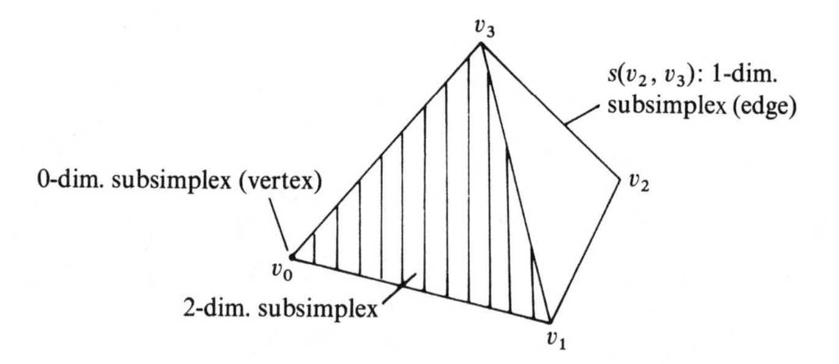

Definition (Simplicial Complex or Polyhedron). A set K of simplices in $\mathbb{R}^n$ is called a simplicial complex or a polyhedron if the following three conditions are satisfied:

(i) If K contains a simplex it contains all faces of this simplex.
(ii) The intersection of two simplices of K is either empty or a common face.
(iii) (In case K is infinite), K is locally finite, i.e. every point of $\mathbb{R}^n$ has a neighborhood that intersects only finitely many simplices of K.

Thus the simplices cannot pierce through one another in a messy way,

but must fit with one another nicely. Here a couple of examples:

(1) Octahedral surface: the polyhedron is composed of eight 2-simplices (and their faces). It is a "simplicial version" of the 2-sphere.

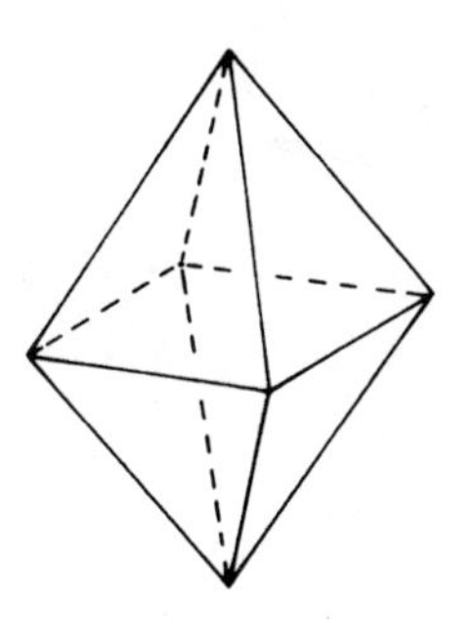

(2) A "simplicial torus":

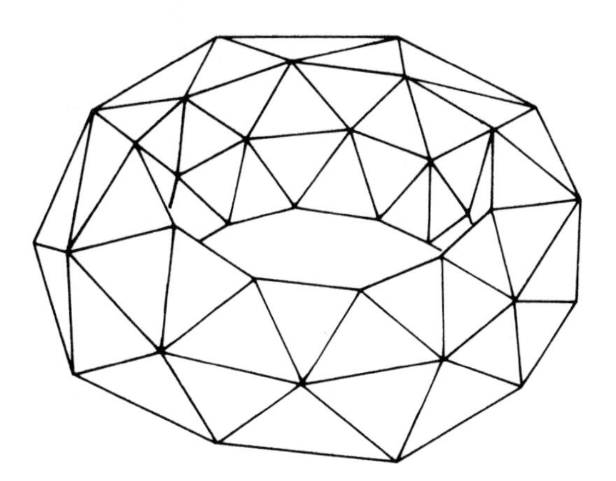

(3) A simplicial Möbius strip:

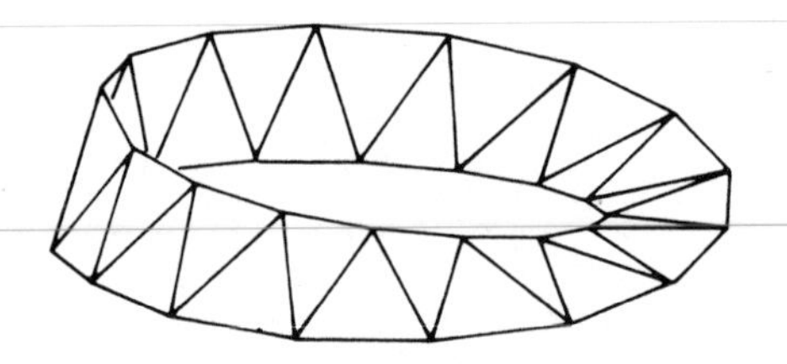

(4) A simplicial something, just as a reminder that simplices can meet in a more general way than in the first three somewhat special examples.

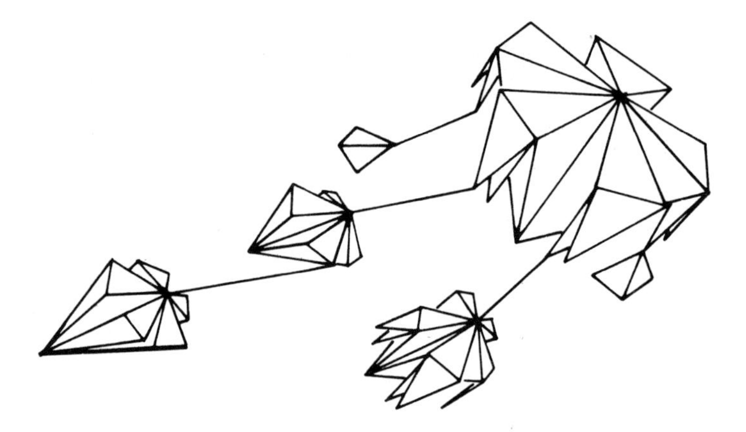

Definition. The subspace $|K| := \bigcup_{s \in K} s$ of $\mathbb{R}^n$ is called the topological space underlying the complex K.

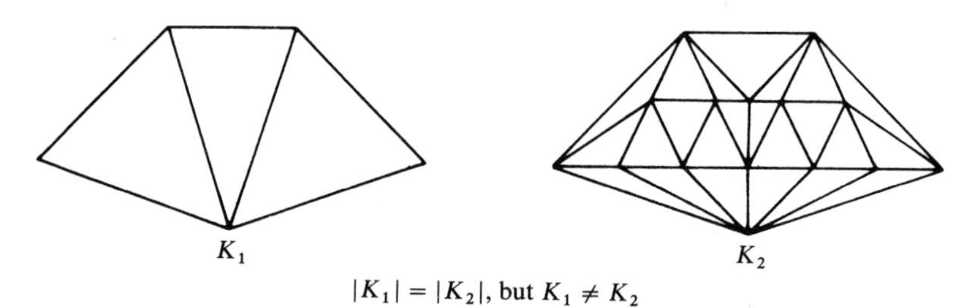

$|K_1| = |K_2|$, but $K_1 \neq K_2$

The difference between K and $|K|$ should be obvious, but you can well imagine that we will not be so pedantic as to stress this difference all the time in notation and words: One talks loosely of a simplicial complex $K \subset \mathbb{R}^n$ (meaning $|K|$), and the minute after one refers to its simplices (now meaning K). Of course there are enough cases where a careful differentiation is essential, as for instance in this chapter.

*

So much for the concept. But what's the use of it? From the topological point of view simplicial complexes, as subspaces of $\mathbb{R}^n$, form only one, apparently very special, class of examples of topological spaces. But with one peculiarity: If for a given complex we know how many essential simplices (i.e. those that are not faces of larger simplices in K) there are in each dimension ("simplex numbers"), and what vertices (hence what sides) these simplices share ("simplex incidences"), then $|K|$ is determined up to a homeomorphism. How do we build from these data a space homeomorphic to $|K|$? We choose for each dimension a standard simplex, for instance $\Delta_k := s(e_1, \ldots, e_{k+1})$ with the unit vectors in $\mathbb{R}^{k+1}$ as vertices; then we form the disjoint sum of as many copies of the standard simplices as are determined by the simplex numbers:

$$X = (\Delta_0 + \cdots + \Delta_0) + \cdots + (\Delta_n + \cdots + \Delta_n)$$

and finally we identify corresponding sides using the incidence data.

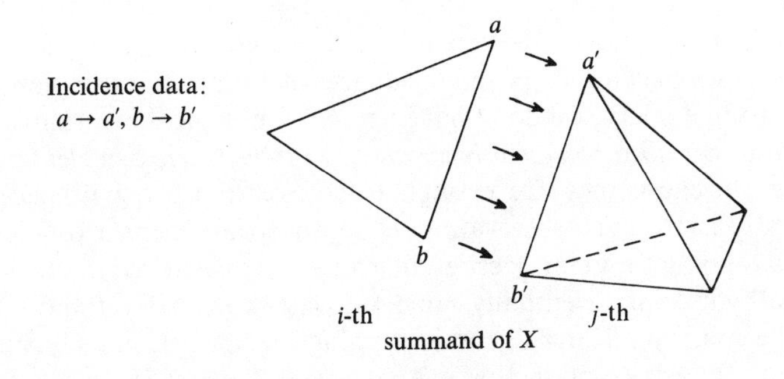

We now have a continuous bijection from the (compact!) quotient space $X/\sim$ to the Hausdorff space $|K|$, which must be a homeomorphism.

Example. Construction of the octahedron from eight 2-simplices:

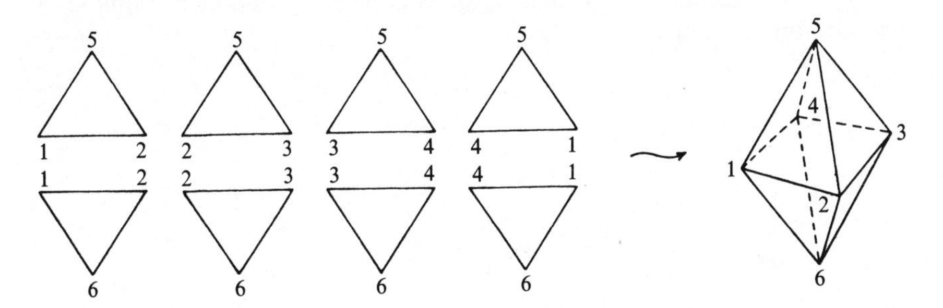

Clearly the simplex numbers and incidences tell us more than the simple homeomorphism type of $|K|$: we know $|K|$ up to a homeomorphism that takes simplices affinely into simplices. But they don't tell us *any more* than that, and in particular notice that the position of $|K|$ in space cannot be established from simplex numbers and incidences, not even "in essence", as shown by the example below:

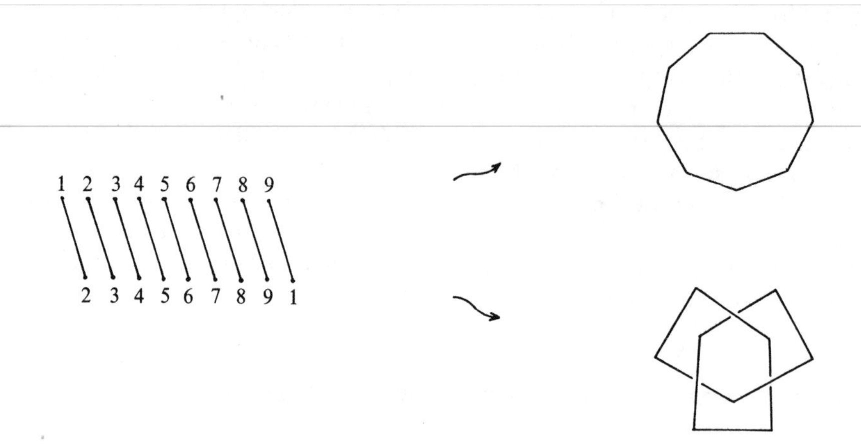

But let's switch back to our main line. If we go over from a topological space to the simplex numbers and incidences of a simplicial complex homeomorphic to that space, we don't have yet any topological invariants, but *we can be sure that all topological invariants can in principle be calculated from these data*, because they are enough to recover the original space up to homeomorphism. This observation is so to speak the starting point of algebraic topology, and for decades all efforts were canalized in the direction indicated by it. What eventually came out was, expressed in today's terminology, the first significant algebraic topological functor, namely simplicial homology. By construction this is a covariant functor $H_* = (H_0, H_1, \ldots)$ from the category of simplicial complexes and simplicial maps (i.e. maps that take k-simplices affinely into k'-simplices $k' \leq k$), or "simplicial category", into the category of graded abelian groups. The decisive factor, however, are invariance theorems which imply that H_* defines a functor (also denoted by H_*) from the category of spaces homeomorphic to simplicial complexes and continuous maps:

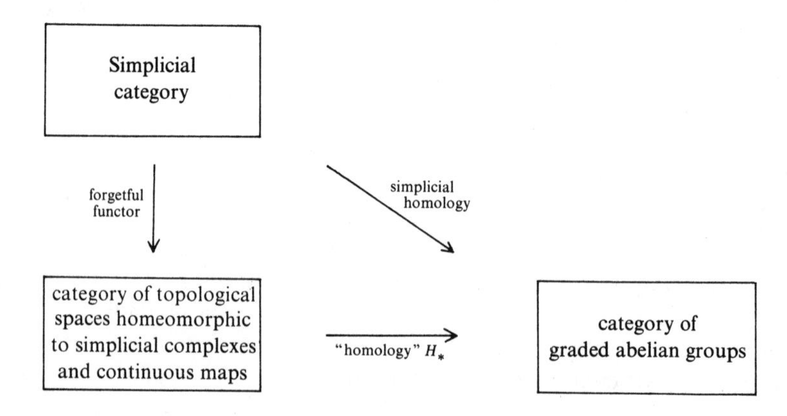

Now even though simplicial homology existed some time before the invariance theorems, it is not to be assumed that the invariance is an "accident" that "luckily" made simplicial homology useful for topology. Clearly the creators of simplicial homology had a geometric intuition for the thing, and from the very beginning had the topological invariance in view.

OK, that was a beginning. Today many other functors have been added, and even homology itself reaches much further now and is expressed in more elegant terms. ("Computing homology with simplicial chains is like computing integrals $\int_a^b f(x)\,dx$ with approximating Riemann sums", A. Dold, *Lectures in Algebraic Topology*, 1972, p. 119.) But building a space out of simple building blocks (here simplices) is, now as before, very useful, the difference being that now one generally uses, instead of simplicial complexes, CW-complexes, which are a sort of "second-generation polyhedra", and much more flexible and practical. What CW-complexes are, their fundamental properties, in what respect they are better than simplicial complexes and why they could be invented only after those, is what I will explain in the next sections.

§2. Cell Decompositions

Just to recall, a *partition* of a set X is a set of pairwise disjoint subsets of X whose union is the whole of X; thus every element of X lies in exactly one such subset. A topological space is called an *n-cell* if it is homeomorphic to $\mathbb{R}^n$; and a *cell decomposition* of a topological space X is a partition of X into subspaces which are cells.

A space with a cell decomposition $(X, \mathscr{E})$ is called a CW-complex if it satisfies certain axioms. I'll enumerate them in the next paragraph; but first let's familiarize ourselves with cells and cell decompositions.

Since $\mathbb{R}^0$ contains only one point, 0-cells are exactly the one-point spaces. The open ball $\mathring{D}^n$ and the punctured sphere $S^n \backslash pt$ are, as is well known, homeomorphic to $\mathbb{R}^n$ and hence n-cells ($S^n \backslash pt \xrightarrow{\cong} \mathbb{R}^n$ by stereographic projection, $\mathbb{R}^n \xleftarrow{\cong} S^n_- \xrightarrow{\cong} \mathring{D}^n$ by central and orthogonal projection).

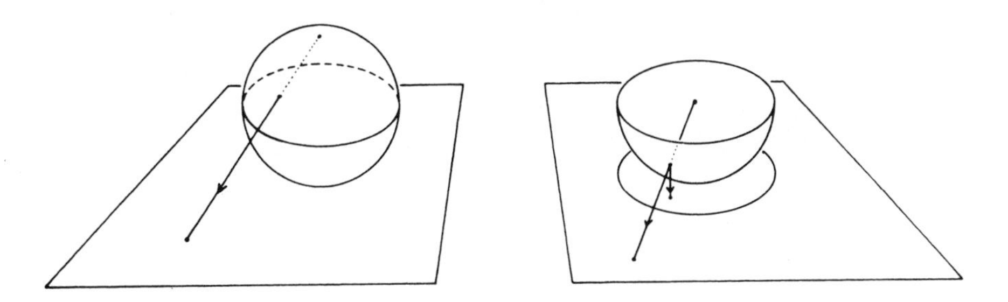

With a positive continuous function $r: S^{n-1} \to \mathbb{R}$ as "stretching factor" we obtain a homeomorphism of $\mathbb{R}^n$ onto itself, given by $0 \mapsto 0$ and

$$x \mapsto r\left(\frac{x}{\|x\|}\right) \cdot x.$$

In particular this homeomorphism takes D^n onto an n-cell, as in the figure:

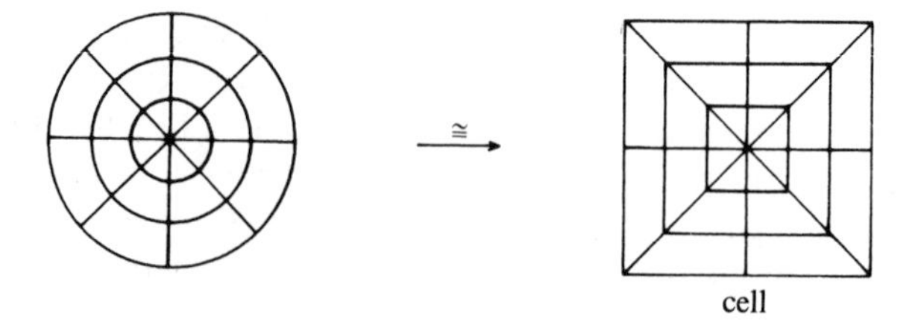

This is a simple procedure, which provides us enough cells. In fact it is even true that every open star-shaped subset of $\mathbb{R}^n$ is an n-cell; the best way to see this is considering the flow of an appropriate radial vector field.

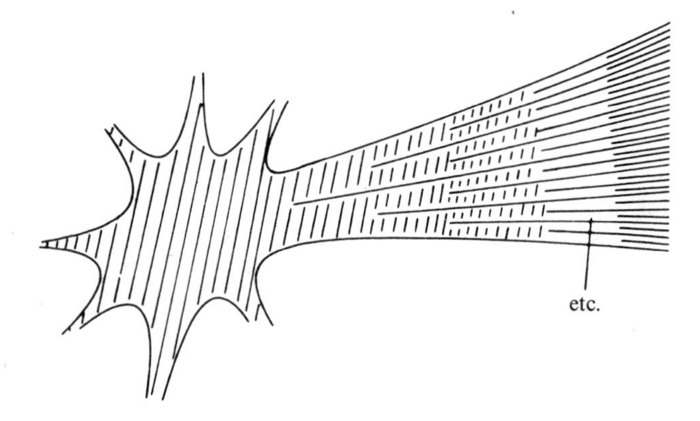

This was just a marginal remark, and I actually shouldn't be diverting your attention to such monstrosities, because although a cell is always a cell, the way the cell in this example lies in $\mathbb{R}^2$ is definitely not typical of the nice and proper way in which cells lie in CW-complexes.

More important is the question of whether an n-cell can at the same time be an m-cell for $m \neq n$. It cannot, I hasten to say. $\mathbb{R}^n \not\cong \mathbb{R}^m$ for $n \neq m$. This was first proved by L. E. J. Brouwer (1911), and the proof is not simple. Only two cases are trivial, $\mathbb{R}^0$ and $\mathbb{R}^1$ are not homeomorphic to any of the higher-dimensional spaces $\mathbb{R}^n$ ($\mathbb{R}^1$ is the only $\mathbb{R}^n$ that can be disconnected by taking away a single point). The proof becomes very simple, though, if one is allowed to apply a bit of algebraic topology: If $\mathbb{R}^n \cong \mathbb{R}^m$, then

$$\mathbb{R}^n \backslash 0 \cong \mathbb{R}^m \backslash 0,$$

thus

$$S^{n-1} \simeq \mathbb{R}^n \backslash 0 \cong \mathbb{R}^m \backslash 0 \simeq S^{m-1},$$

and because of the homotopy invariance of homology it follows that

$$H_{n-1}(S^{n-1}) \cong H_{n-1}(S^{m-1}).$$

But $H_k(S^i) \cong \mathbb{Z}$ for $i = k > 0$ and zero for $i \neq k > 0$, hence $n = m$, qed. By the way, this is really an "honest" proof, because in the derivation of the homological tools the Brouwer theorem is never used. We can thus talk about *the* dimension of a cell.

So much for the cells as individual beings. Now let's look around for some examples of cell decompositions. Every simplicial complex K defines in a canonical way a cell decomposition of $|K|$, as follows: The union of the proper faces of a simplex s is called its boundary ∂s, and $s \backslash \partial s$ is called the "open simplex" associated with s. Open simplices are cells, and the collection of open simplices of a simplicial complex K forms a cell decomposition for $|K|$. A couple of other examples:

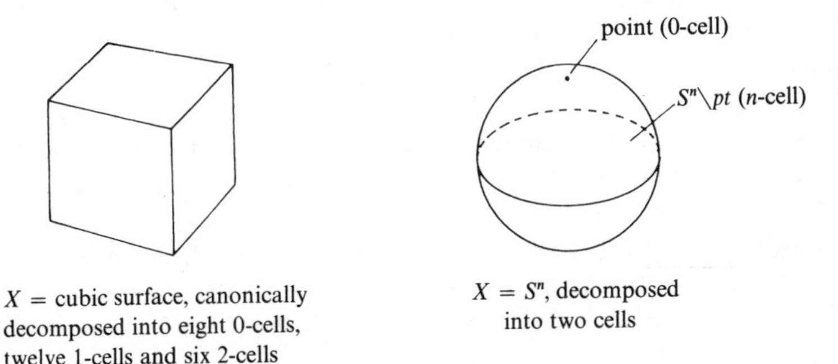

X = cubic surface, canonically decomposed into eight 0-cells, twelve 1-cells and six 2-cells

$X = S^n$, decomposed into two cells

These are very well-behaved examples. Unhindered by axioms we could of course decompose a space by choosing for instance some very wild pairwise disjoint cells contained in it (like the "star" in the preceding page), and the remaining points would be defined as 0-cells of the decomposition. But with such decompositions it is impossible to start anything reasonable, and that's why we now turn to the "CW-axioms".

§3. The Notion of a CW-Complex

Definition (CW-Complex). A pair $(X, \mathscr{E})$, consisting of a Hausdorff space X and a cell decomposition $\mathscr{E}$ of X, is called a *CW-complex* if the following three axioms are satisfied:

Axiom 1 ("Characteristic Maps"). For each n-cell $e \in \mathscr{E}$ there is a continuous map $\Phi_e \colon D^n \to X$ taking $\mathring{D}^n$ homeomorphically onto the cell e and S^{n-1} into the union of the cells of dimension at most $n - 1$.

Axiom 2 ("Closure Finiteness"). The closure $\bar{e}$ of each cell $e \in \mathscr{E}$ intersects only a finite number of other cells.

Axiom 3 ("Weak Topology"). $A \subset X$ is closed if and only if every $A \cap \bar{e}$ is.

This concept was introduced in 1949 by J. H. C. Whitehead, and the name CW-complex refers to Axioms 2 and 3, which give conditions under which *infinitely many* cells are allowed in the complex (for finite cell decompositions these two axioms are always trivially satisfied): "C" stands for "closure finiteness" and "W" for "weak topology".

Definition. If X is a space decomposed into cells, X^n denotes the union of cells of dimension $\leq n$, and is called the n-skeleton of X.

Axiom 1 (about the existence of characteristic maps) says roughly that the n-cells are to be thought of as "attached" to the $(n-1)$-skeleton. Later we'll make this statement more precise (§5). Before considering examples to illustrate the three axioms, I would like to mention a coupled of immediate consequences of Axiom 1, which should be included right after the definition in your intuition about CW-complexes. For instance, every non-empty CW-complex must contain at least one 0-cell, for if $n > 0$ were the lowest dimension of a cell, S^{n-1} $(\neq \varnothing)$ could not be taken into $X^{n-1} \neq \varnothing$. It also follows at once that every finite CW-complex is compact, being the union of finitely many compact subspaces $\Phi_e(D^n)$, $e \in \mathscr{E}$. It is actually true even that the closure of every cell is compact. More precisely:

Proposition. *If a cell decomposition of a Hausdorff space X satisfies Axiom 1, then for every n-cell we have $\bar{e} = \Phi_e(D^n)$ and in particular the closure $\bar{e}$ is compact and the "cell boundary" $\bar{e} \backslash e = \Phi_e(S^{n-1})$ lies in the $(n-1)$-skeleton.*

PROOF. It is true in general for continuous maps that $f(\bar{B}) \subset \overline{f(B)}$; thus $\bar{e} = \overline{\Phi_e(\mathring{D}^n)} \supset \Phi_e(D^n) \supset e$. Being a compact subspace of a Hausdorff space, $\Phi_e(D^n)$ is closed, and thus equal to $\bar{e}$ because it lies between e and $\bar{e}$. qed. □

Now let's take a look at some examples of cell decompositions of Hausdorff spaces, with an eye towards the axioms:

First some finite decompositions, in which case Axioms 2 and 3 are automatically satisfied.

(1)

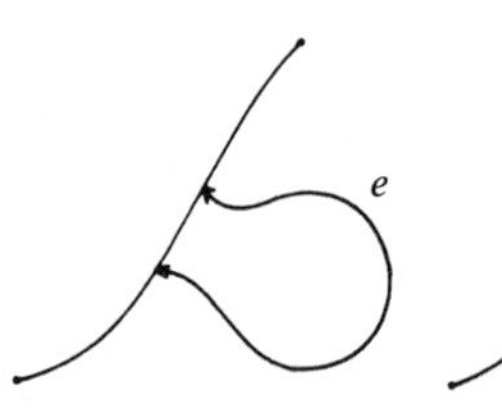

Two 0-cells, two 1-cells Axiom 1 fails, cell boundary of e is not in 0-skeleton

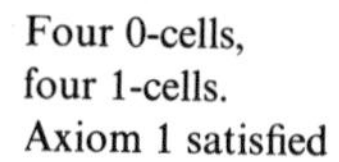

Four 0-cells, four 1-cells. Axiom 1 satisfied

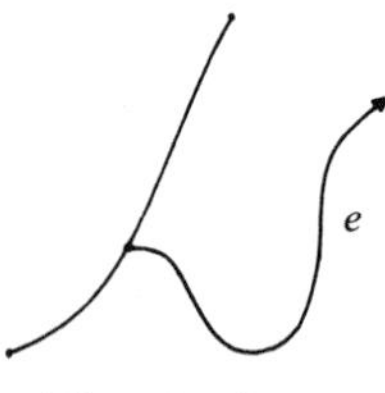

Three 0-cells, three 1-cells. Axiom 1 fails, cell closure e is not compact.

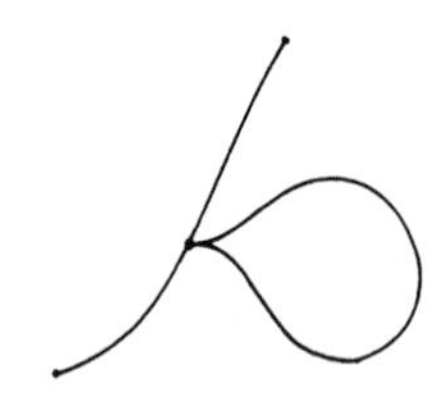

Three 0-cells, three 1-cells. Axiom 1 satisfied.

(2) The following decomposition into three 0- and two 1-dimensional cells does not satisfy Axiom 1, since the boundary of cell e is not in the 0-skeleton. Incidentally, the example cannot be "fixed" by considering some other decomposition: the space is not CW-decomposable.

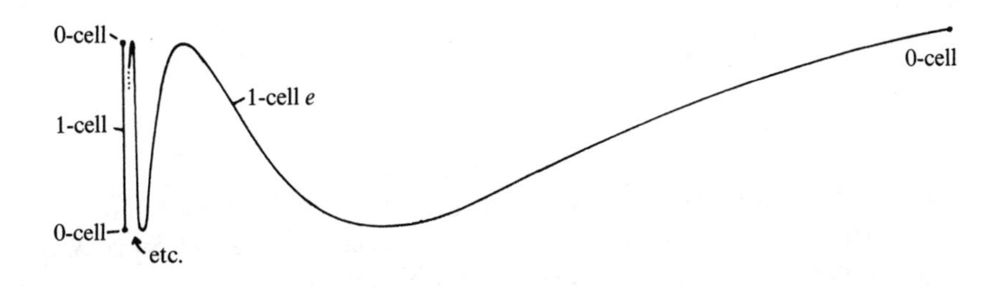

(3) The decompositions of the cube and the sphere given at the end of the preceding paragraph are CW-decompositions.

Now two examples demonstrating the independence of Axioms 2 and 3:

(4)

(5)

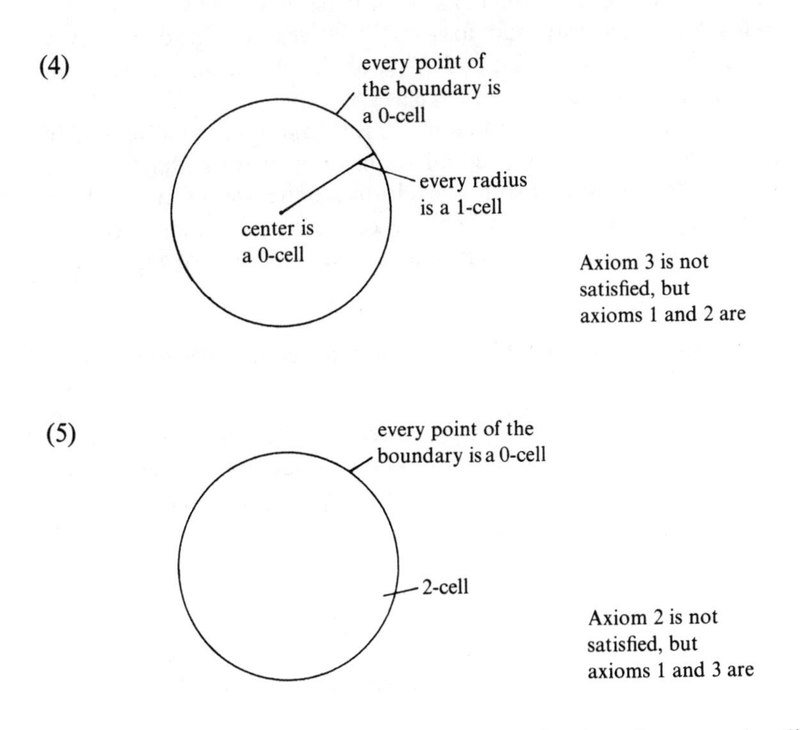

(6) The decomposition of a simplicial complex into its open simplices is a CW-decomposition.

§4. Subcomplexes

Definition and Lemma (Subcomplexes). Let $(X, \mathscr{E})$ be a CW-complex, $\mathscr{E}' \subset \mathscr{E}$ a set of cells in it, and $X' = \bigcup_{e \in \mathscr{E}'} e$ its union. $(X', \mathscr{E}')$ is called a subcomplex of $(X, \mathscr{E})$ if one of the following three equivalent conditions is satisfied:

(a) $(X', \mathscr{E}')$ is also a CW-complex;
(b) X' is closed in X;
(c) $\bar{e} \subset X'$ for every $e \in \mathscr{E}'$.

PROOF OF THE EQUIVALENCE OF THE THREE CONDITIONS. (b) $\Rightarrow$ (c) is trivial.
(c) $\Rightarrow$ (b). We have to show that $\bar{e} \cap X'$ is closed for all $e \in \mathscr{E}$. Since X is closure finite $\bar{e} \cap X' = \bar{e} \cap (e_1' \cup \cdots \cup e_r')$, which by (c) equals

$$\bar{e} \cap (\bar{e}_1' \cup \cdots \cup \bar{e}_r'),$$

being thus closed, qed.
(a) $\Rightarrow$ (c). A characteristic map Φ_e for $e \in \mathscr{E}'$ relative to $(X', \mathscr{E}')$ is also characteristic relative to $(X, \mathscr{E})$, hence the remark in §3 implies that $\Phi_e(D^n)$, the closure of e in X (which is of course what we mean in (c)) is also the closure of e in the subspace X', and consequently lies in it. qed.
(b, c) $\Rightarrow$ (a). A characteristic map for $e \in \mathscr{E}'$ relative to X is also characteristic relative to X', because of (c); and X' is obviously closure finite. Thus $(X', \mathscr{E}')$ satisfies Axioms 1, 2. We still have to show that if $A \subset X'$ and $A \cap \bar{e}$ is closed in X' for all $e \in \mathscr{E}'$, A is closed in X'. But (b) implies that "closed in X" is the same as "closed in X'", so all we have to prove is that $A \cap \bar{e}$ is closed for $e \in \mathscr{E} \backslash \mathscr{E}'$ as well. From the closure finiteness of X we have $A \cap \bar{e} = A \cap (e_1' \cup \cdots \cup e_r') \cap \bar{e}$, where we can take $e_i' \in \mathscr{E}'$, since cells in $\mathscr{E} \backslash \mathscr{E}'$ can't contribute to the intersection with $A \subset X'$. Then we get

$$A \cap \bar{e} = A \cap (\bar{e}_1' \cup \cdots \cup \bar{e}_r') \cap \bar{e},$$

but $A \cap (\bar{e}_1' \cup \cdots \cup \bar{e}_r')$ is closed by assumption, hence $A \cap \bar{e}$ is too. qed.

□

It is fair to say that from this easily remembered lemma there follows immediately everything that one wants to know about subcomplexes when dealing practically with CW-complexes. Let's enumerate a couple of such consequences:

Corollaries.

(1) *Arbitrary intersections* (*using* (b)) *and also arbitrary unions* (*using* (c)) *of subcomplexes are again subcomplexes.*
(2) *The skeletons are subcomplexes* (*using* (c) *and the proposition in* §3).
(3) *Every union of n-cells in* $\mathscr{E}$ *with* X^{n-1} *forms a subcomplex* (*same reason*).
(4) *Every cell lies in a finite subcomplex* (*induction on dimension of cell*: *closure finiteness and proposition in* §3).

A fifth consequence will be stated separately because it deserves more attention:

Corollary. *Every compact subset of a* CW-*complex is contained in a finite subcomplex. In particular a* CW-*complex is compact* if and only if *it is finite.*

PROOF. Using (1) and (4) there remains to prove that a compact subset $A \subset X$ meets only a finite number of cells. To prove this, choose a point in each cell met by A. This set of points P is closed, because closure finiteness implies that $P \cap \bar{e}$ is actually finite and we're in a Hausdorff space. But this argument works for any subset of P! Thus P has the discrete topology, but as a closed subset of a compact A it must be compact too, hence finite, qed. □

§5. Cell Attaching

So far we have talked about CW-complexes as existing objects whose properties we are studying. Now I will indicate the most important method for the *construction* of CW-complexes. It is easy to visualize, being essentially the cell attaching that we have already considered in III, §7, Example 1. This method is not only of practical but also of fundamental importance, because since *every* CW-complex, up to cell-preserving homeomorphisms, can be so represented, one gets in this way a certain overall understanding about the possible CW-complexes. I will not include the proof, but it is not difficult and does not depend on anything not dealt with in this book (cf. III, §§1–3, 7).

If X is a CW-complex and $\varphi: S^{n-1} \to X^{n-1}$ is a continuous map into the $(n-1)$-skeleton, $X \cup_\varphi D^n$ is again a CW-complex with one more cell, in a canonical way. The canonical map $D^n \subset X + D^n \to X \cup_\varphi D^n$ is characteristic. The cell boundary of the new cell is $\varphi(S^{n-1}) \subset X^{n-1}$. Note that this cell boundary doesn't have to be a homeomorphic image of the sphere, just a continuous image.

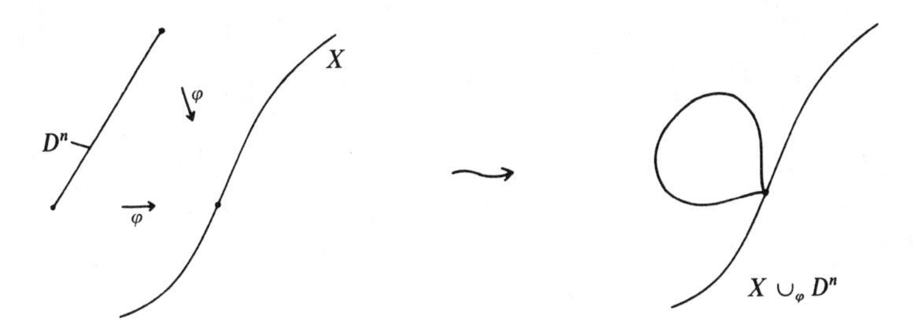

Analogously, we can attach a whole family of n-cells simultaneously: Let $\{\varphi_\lambda\}_{\lambda \in \Lambda}$ be a family of continuous maps $\varphi_\lambda: S^{n-1} \to X^{n-1}$, and let's collect them together into a single continuous map $\varphi: S^{n-1} \times \Lambda \to X^{n-1}$,

$(v, \lambda) \mapsto \varphi_\lambda(v)$, where Λ has the discrete topology. Then $X \cup_\varphi (D^n \times \Lambda)$ is again a CW-complex in a canonical way, created from X by "attaching a family of n-cells". Notice that by no means the boundaries of the new cells have to be disjoint:

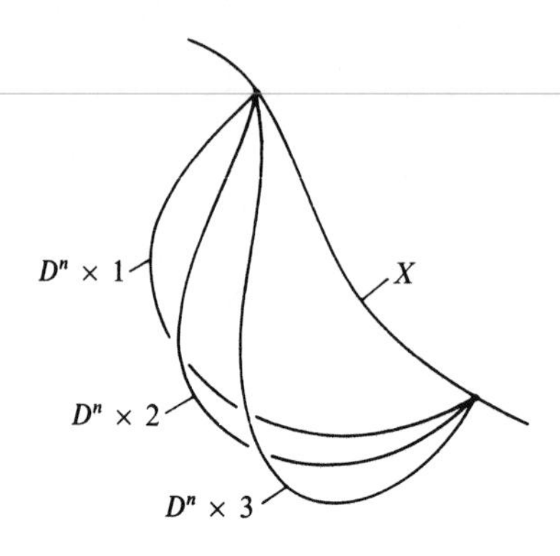

But now we can obtain any CW-complex by successively attaching families of cells: We start with the zero-skeleton X^0. This is simply a discrete space, and if we wish we can think of it as being obtained by attaching a family of 0-cells to the empty space. How do we get X^n from X^{n-1}? Let $\mathscr{E}^n$ be the set of n-cells. Choose for every n-cell e a characteristic map Φ_e and set $\varphi_e := \Phi_e | S^{n-1}$. If we now use $\{\varphi_e\}_{e \in \mathscr{E}^n}$ as the family of attaching maps, we get after the attaching a CW-complex $X^{n-1} \cup_\varphi (D^n \times \mathscr{E}^n)$, which is cell-preserving homeomorphic to X^n in a canonical way.

We thus get all skeletons by induction, and in particular X itself if X is finite-dimensional (i.e. contains no cell above a certain dimension). On the other hand, if X is infinite-dimensional we get X from the skeletons

$$X^0 \subset X^1 \subset \cdots$$

as the union $\bigcup_{n=0}^{\infty} X^n$, endowed with the "weak topology" prescribed by Axiom 3.

§6. Why CW-Complexes Are More Flexible

I'll now mention some points of view from which CW-complexes are "more well behaved" or "more convenient" then simplicial complexes. Let's start with Cartesian products. The product of two cells is, of course, another cell, and if $(X, \mathscr{E})$ and $(Y, \mathscr{F})$ are cell decompositions for X and Y respectively, then $\{e \times e' | e \in \mathscr{E}, e' \in \mathscr{F}\}$ is a cell decomposition for $X \times Y$, and it is easy to prove that the following holds for such decompositions:

Note. *If X and Y are finite* CW-*complexes, then $X \times Y$ is also a* CW-*complex.*

Remark. (no proof here, see for instance Dold [5], p. 99). For infinite CW-complexes it can happen that $X \times Y$ does not have the weak topology (but Axioms 1 and 2 are always satisfied). But under very mild additional assumptions, for instance when one of the factors is locally compact, $X \times Y$ is again a CW-complex.

The product of two positive-dimensional simplices, though, is not a simplex anymore:

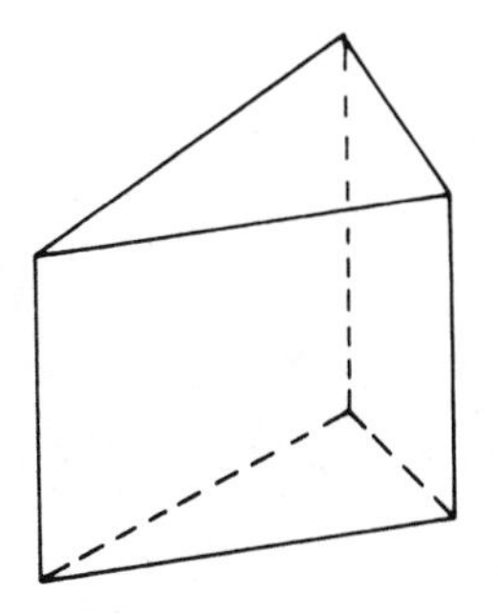

so that if we want to make the product of two simplicial complexes into one, we have to subdivide further the individual simplex products.

*

In III, §6 we had considered a number of examples of "collapsing" a subspace to a point, and it is exactly in algebraic topology that this operation comes up most often. For CW-complexes the following is very easy to verify:

Note. *If X is a* CW-*complex and $A \subset X$ a subcomplex the cell decomposition of X/A into the* 0-*cell A and the cells of $X \setminus A$ is again a* CW-*decomposition, or, in short: X/A is a* CW-*complex in a canonical way* (Dold [5], p. 98).

For simplicial complexes, on the contrary, there is no such canonical quotient operation. The quotient X/A of a simplicial complex by a subcomplex cannot in general be made simplicial without further subdivision and a new "embedding" in a possibly much higher-dimensional Euclidean space. Just think as a simple example that the quotient of a single simplex by its boundary is homeomorphic to the sphere.

Analogously, there is no "simplex attaching" to correspond to cell attaching in CW-complexes; already when we simply glue together two 1-simplices by the boundary, we have to do some work and make choices to get a simplicial complex homeomorphic to the result (here S^1).

*

For a CW-decomposition of a space X much fewer and "more natural" cells are generally necessary than for a simplicial complex homeomorphic to X. To exemplify let's consider some simple spaces.

(1) The sphere S^2 as a simplicial and a CW-complex:

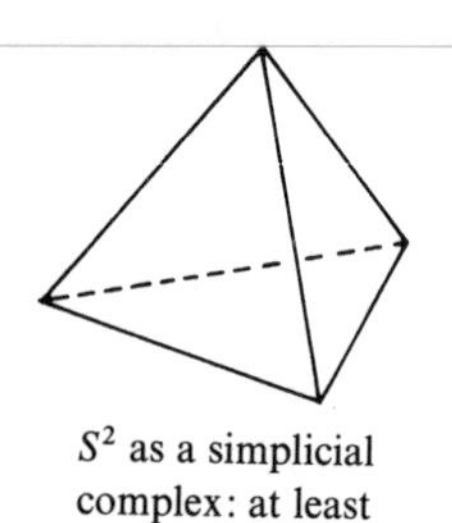

S^2 as a simplicial complex: at least 14 simplices

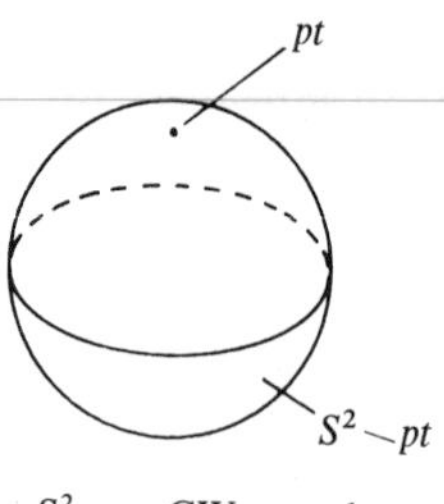

S^2 as a CW-complex: it works with 2 cells

(2) The torus $S^1 \times S^1$. Since S^1 can be CW-decomposed into two cells, we can do it for $S^1 \times S^1$ with four:

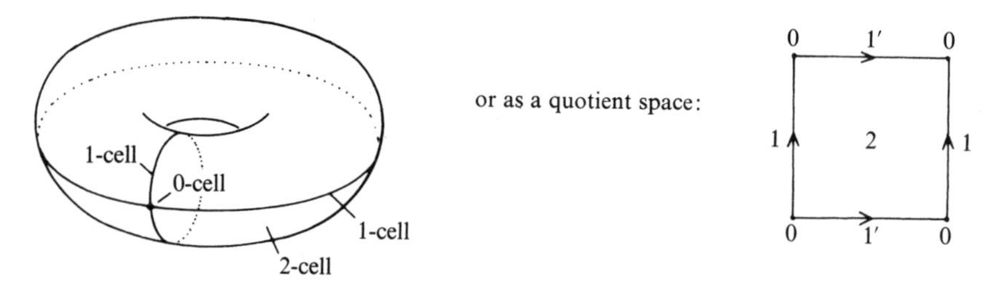

In contrast, to build a simplicial complex homeomorphic to a torus, we need quite a few simplices—42, to be exact, as Professor Guy Hirsch casually informed me over a cup of tea, correcting a rash conjecture of mine.

(3) The n-dimensional projective space can be CW-decomposed into $n + 1$ cells in an entirely natural way:

$$\mathbb{RP}^n = e_0 \cup \cdots \cup e_n,$$

$$\mathbb{CP}^n = e_0 \cup e_2 \cup \cdots \cup e_{2n},$$

where the cells are the affine spaces $\mathbb{P}^n = \mathbb{P}^0 \cup (\mathbb{P}^1 \backslash \mathbb{P}^0) \cdots \cup (\mathbb{P}^n \backslash \mathbb{P}^{n-1})$. There is no simplicial decomposition as simple and natural.

§7. Yes, But . . . ?

This is all nice and well, but simplicial complexes were not being considered for their own sake, but because they accomplish something: the algebraization of geometric objects, the calculation of the homology functor and the topological invariants connected with it. . . . What do CW-decompositions accomplish, convenient as they may be? The question is perfectly warranted.

Comparing what we do in both cases to construct a space from the building blocks,

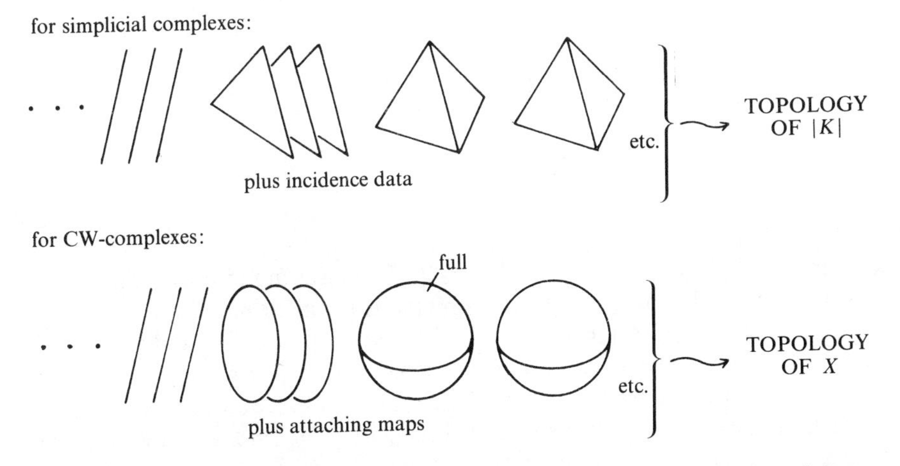

we immediately notice a big difference relevant to the possible use of these notions: while for simplicial complexes the incidence data are something algebraic and thus establish some sort of primitive "algebraization" of $|K|$, the attaching maps are only continuous maps $\varphi: S^{n-1} \to X^{n-1}$, and thus complicated geometrical objects, themselves needing algebraization. It is thus not immediately clear what we stand to gain by substituting cells and attaching maps for a space. This is also the reason why, even assuming that CW- and simplicial complexes were invented around the same time, the latter must have been preferred.

And now we come to the crux of the matter: The investigation of simplicial complexes led to the development of homology theory, and homology theory itself can be used in the algebraization of attaching maps. The homological properties of attaching maps (as I must say somewhat vaguely, since getting into homology in more detail would lead us too far afield) can be expressed by means of certain "incidence numbers". Such numbers do not contain full information about the attaching maps anymore, nor do they allow us to recover fully the topology of the complex. But they are enough to determine the homology of the complex,

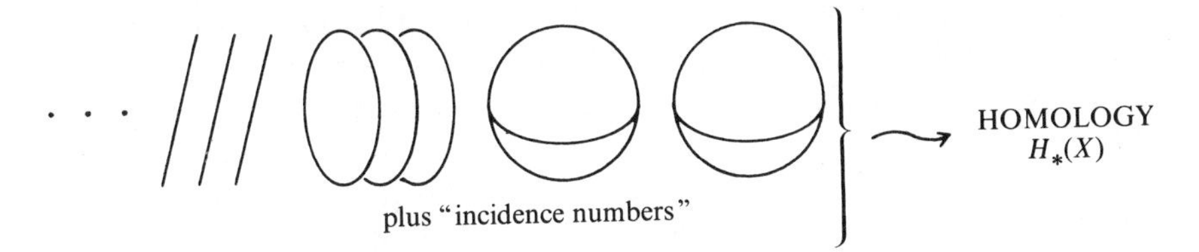

and the point is that this method is extraordinarily more effective and faster than the direct calculation of simplicial homology.

Having said this much, I should also indicate, though without proof and further elucidation, a useful and easily remembered consequence: The Euler characteristic of a finite CW-complex is the alternate sum of the number of cells in each dimension. So we obtain in our examples, for instance:

$$\chi(S^n) = 1 + (-1)^n,$$

$$\chi(S^1 \times S^1) = 1 - 2 + 1 = 0,$$

$$\chi(\mathbb{RP}^n) = \tfrac{1}{2}(1 + (-1)^n),$$

$$\chi(\mathbb{CP}^n) = n + 1.$$

To conclude let me mention that CW-complexes are important on other grounds as well. For example, a geometrical problem often boils down to finding a continuous map $f: X \to Y$ with certain properties, and frequently a CW-decomposition of X is what is needed to shed the right light on things. Because then one tries to construct the map by induction on the skeleton. The first map $f_0: X^0 \to Y$ is generally easy enough to find, and if we already have

$$f_{n-1}: X^{n-1} \to Y,$$

a continuous map $f_{n-1} \circ \Phi_e | S^{n-1}: S^{n-1} \to Y$ is given for each n-cell e,

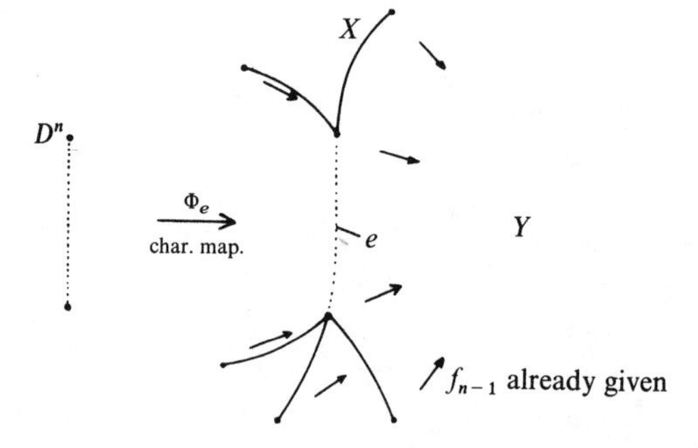

and it follows easily from the axioms for CW-complexes that f_{n-1} can be extended to a continuous $f_n: X^n \to Y$ if and only if each one of these maps $\alpha_e := f_{n-1} \circ \Phi_e | S^{n-1}$ can be extended to a continuous map $D^n \to Y$. This in turn means that the element $[\alpha_e] \in \pi_{n-1}(Y)$ must be zero, which is certainly the case for instance when this homotopy group is trivial . . . and so on.

"Simplifying assumptions" make life easier for the mathematician, but when can they be made? In algebraic topology a compromise often has to be worked out: The spaces must be special enough to allow certain methods to work and certain theorems to apply, but at the same time they must be general enough to include certain important examples of applications. Considering CW-complexes (or spaces homotopy equivalent to them) is often a good compromise of this sort, and this is another reason to be acquainted with the notion of a CW-complex.

CHAPTER VIII

Construction of Continuous Functions on Topological Spaces

§1. The Urysohn Lemma

When we want to construct a function with certain properties on $\mathbb{R}^n$ or one of its subspaces, analysis offers us a vast array of tools. First thing to come to mind are perhaps polynomials and rational functions, and there is a lot that can be done with them alone! Then come the so called "elementary functions", like the exponential, the logarithm, the trigonometric functions; then power series or, more generally, uniformly convergent sequences of already available continuous functions. We can obtain functions with given properties as solutions of differential equations, and so on and on—as we might say in good conscience.

All this looks a bit more difficult on manifolds, but the ties between manifolds and analysis are so close that we still have essentially the same rich possibilities for the construction of continuous functions. For one thing, many analytical techniques (e.g. differential equations) can be carried over to differentiable manifolds; then we can also imbed manifolds in $\mathbb{R}^N$

$$(M^n \cong M' \subset \mathbb{R}^N, \text{ for } N \text{ big enough}),$$

where they can be treated as subsets of $\mathbb{R}^N$; and finally we have a third alternative, which is often the most practical: we establish a direct relation between the manifold and $\mathbb{R}^N$ by means of charts.

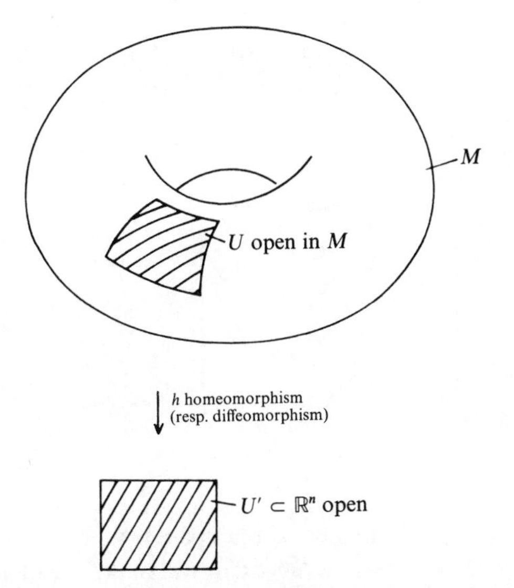

and we can then "lift" a continuous function f on U' to a continuous function on U

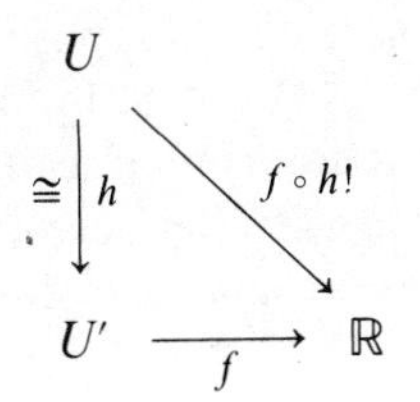

Granted that a function on U is not a function on the whole of M, but if for instance the *support* of the function f to be lifted (i.e. the closure $\operatorname{Supp} f := \overline{\{x \mid f(x) \neq 0\}}$) is *compact*,

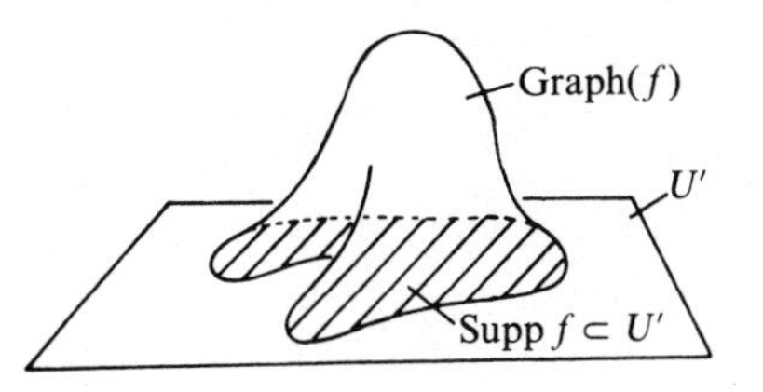

the composition $f \circ h$ can simply be extended to a continuous function F on the whole of M taking the value 0 outside U:

$$F(p) := \begin{cases} f \circ h(p) & \text{for } p \in U, \\ 0 & \text{otherwise.} \end{cases}$$

Such functions then play an important role, either achieving themselves the end in view, or serving as a means to achieve it ("partitions of unity", §4).

CW-complexes also have a similar, if not so close, connection to analysis. Here, using induction over the skeletons, we face the problem of extending an existing function on S^{n-1} to the whole disk D^n, taking into account the desired properties.

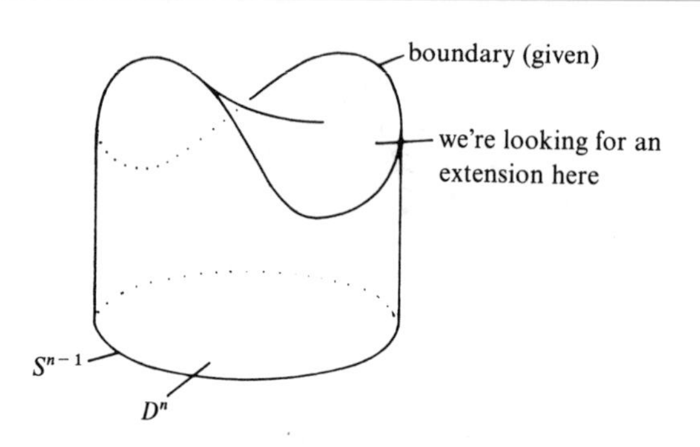

And finally let's mention metrizable spaces, whose structure is already considerably poorer: But even here we can still make use of a metric

$$d: X \times X \to \mathbb{R},$$

whenever we need a function on X. Suppose for instance the problem is finding a continuous function $f: X \to [0, 1]$ for a given neighborhood of a point $p \in X$, such that the function is identically zero outside this neighborhood and identically one inside a smaller neighborhood. Then we just choose $0 < \varepsilon < \delta$ small enough, take an auxiliary function $\lambda: \mathbb{R} \to [0, 1]$ looking like this

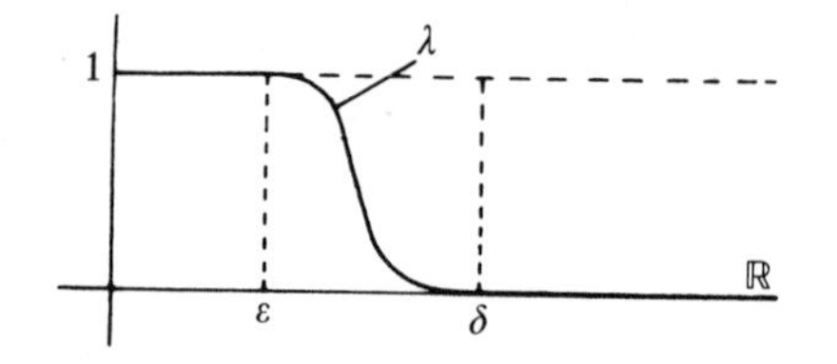

and make $f(x) := \lambda(d(x, p))$.

This introduction was just to make sure that you really see the problem of constructing continuous functions on general topological spaces as a problem. Imagine you are given $V \subset U \subset X$ and you have to find a function

$$f: X \to [0, 1]$$

that is identically one on V and identically zero outside U.

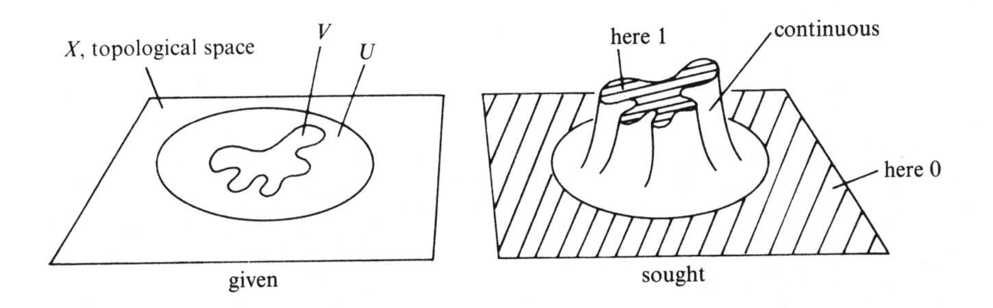

Where are we going to get such a continuous transition from 1 to 0, if the topological space X has no recognizable relation to the real numbers—no charts, no cells, no metric? So this is the problem:

Fundamental problem in the construction of functions on topological spaces (The Urysohn Lemma Problem). Let A and B be closed, disjoint subsets of a topological space X. Find a continuous function $f : X \to [0, 1]$ with $f|A \equiv 1$ $f|B \equiv 0$.

Notice that for every continuous function $f : X \to \mathbb{R}$ the sets $f^{-1}(1)$ and $f^{-1}(0)$ are closed anyway, so that the problem above is solvable for arbitrary A and B if and only if it is solvable for $\bar{A}$ and $\bar{B}$: that's why we only have to consider the case when A and B are closed.

A *necessary* condition for the problem to have a solution can be immediately stated: A and B must be separable by open neighborhoods, because if there is such an f, the sets $f^{-1}(\frac{3}{4}, 1]$ and $f^{-1}[0, \frac{1}{4})$, for instance, form open, disjoint neighborhoods of A and B. (Here and in the following we use the common expression "open neighborhood" U of a subset $A \subset X$, meaning simply an open set U containing A.)

The existence of separating open neighborhoods for A and B is not sufficient, but we have the following result, which is so to speak the fundamental theorem for the construction of functions on topological spaces:

Urysohn Lemma. *Suppose for the topological space* X every *pair of disjoint closed sets can be separated by disjoint open neighborhoods. Then for every pair of disjoint closed sets there is a continuous function* $f : X \to [0, 1]$ *which takes the value* 1 *on one set and the value* 0 *on the other.*

The proof comes in §2. But before we go into it we will take a look at some classes of topological spaces with this separation property. First remark that all metric spaces satisfy it trivially: Let (X, d) be a metric space. Given a non-empty closed set B, a point $a \notin B$ has always a positive "distance" $\inf_{x \in B} d(a, x) > 0$ to B, because there must be a whole ball around a lying outside B. Denote by $U_B(a)$ the open ball around a with radius *half* the distance to B.

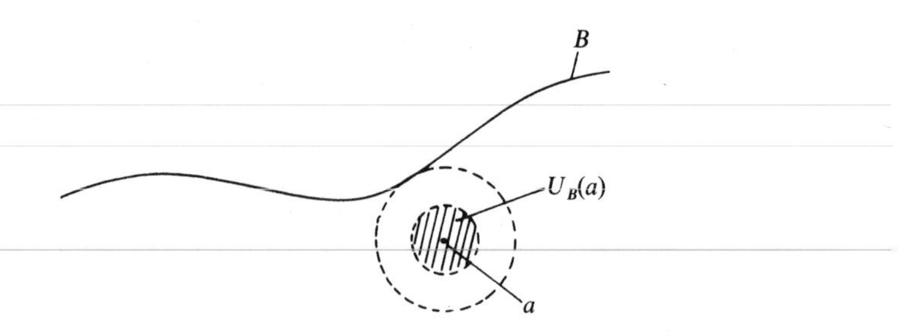

Now let A and B be disjoint closed sets, we get separating open neighborhoods simply by putting $U := \bigcup_{a\in A} U_B(a)$ and $V := \bigcup_{b\in B} U_A(b)$.

In CW-complexes, too, it is always possible to separate closed sets by open neighborhoods: using induction over the skeletons one reduces the question to an easily solvable problem in D^n.

For a third example, notice the following

Proposition. *In every compact Hausdorff space two closed disjoint sets can be separated by open neighborhoods.*

PROOF. Every two points $a \in A$ and $b \in B$ can be separated by open neighborhoods $U(a, b)$ and $V(a, b)$ because of Hausdorffness. For a fixed, find $b_1, \ldots, b_r \in B$ with $B \subset V(a, b_1) \cup \cdots \cup V(a, b_r)$, possible because B is a closed subspace of a compact space, hence compact. Then

$$U(a) := U(a, b_1) \cap \cdots \cap U(a, b_r)$$

and

$$V(a) := V(a, b_1) \cup \cdots \cup V(a, b_r)$$

are separating neighborhoods for a and B

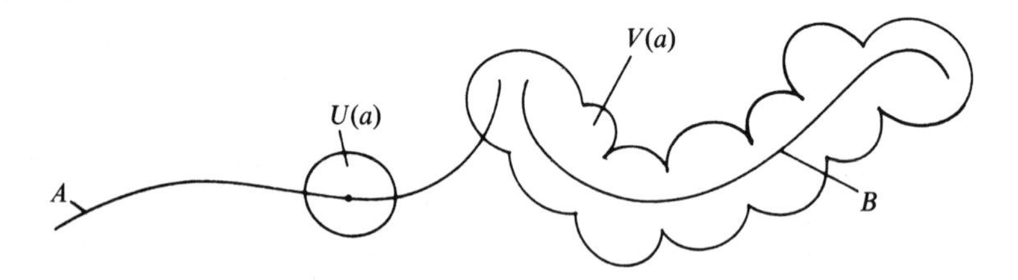

and, analogously, $U := U(a_1) \cap \cdots \cap U(a_s)$ and $V := V(a_1) \cup \cdots \cup V(a_s)$ separate A and B, qed. □

Thus the Urysohn lemma is also applicable to compact Hausdorff spaces, and about such spaces it cannot be said that they were "*a priori*" or "by definition" related to the real numbers. So you must concede that the Urysohn lemma is indeed a striking theorem.

But maybe you'll want to revise this opinion about the lemma after seeing the proof. It is quite simple and may leave you with the feeling that "I could have thought of it myself". But isn't this a bit of self-deception? Just try it before reading the proof....

§2. The Proof of the Urysohn Lemma

The simple idea is to obtain the function as a limit of step functions that decrease gradually from A and B and get ever finer:

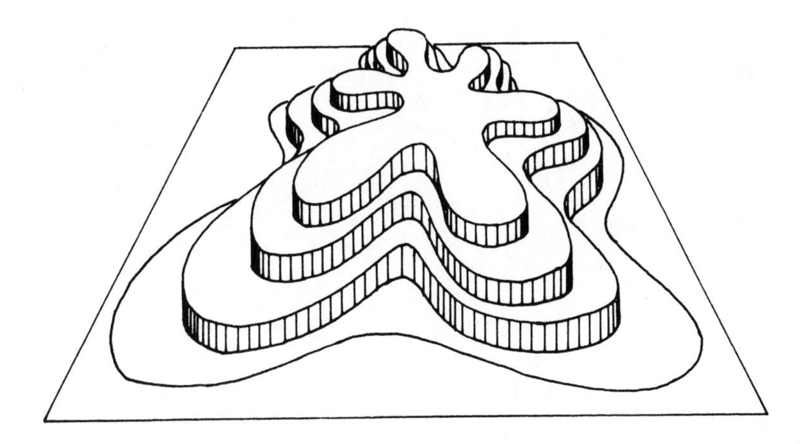

Giving such a step function is simply the same as giving a chain of sets "between" A and $X \backslash B$:

$$A = A_0 \subset A_1 \subset \cdots \subset A_n \subset X \backslash B.$$

Then we define the "corresponding" step function as being equal to 1 on A_0, to $1 - 1/n$ on $A_1 \backslash A_0$, to $1 - 2/n$ on $A_2 \backslash A_1$ etc., and equal to zero outside A_n, and in particular on B. Such a function is of course not continuous. In order to make its "jumps" successively smaller and smaller, so as to obtain in the limit a continuous function, we are going to indent each step with additional "plateaus", halving the jump height each time:

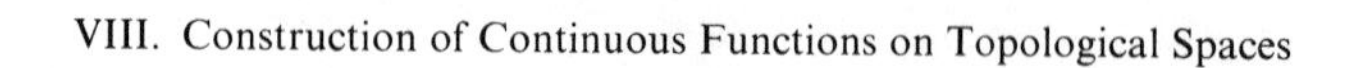

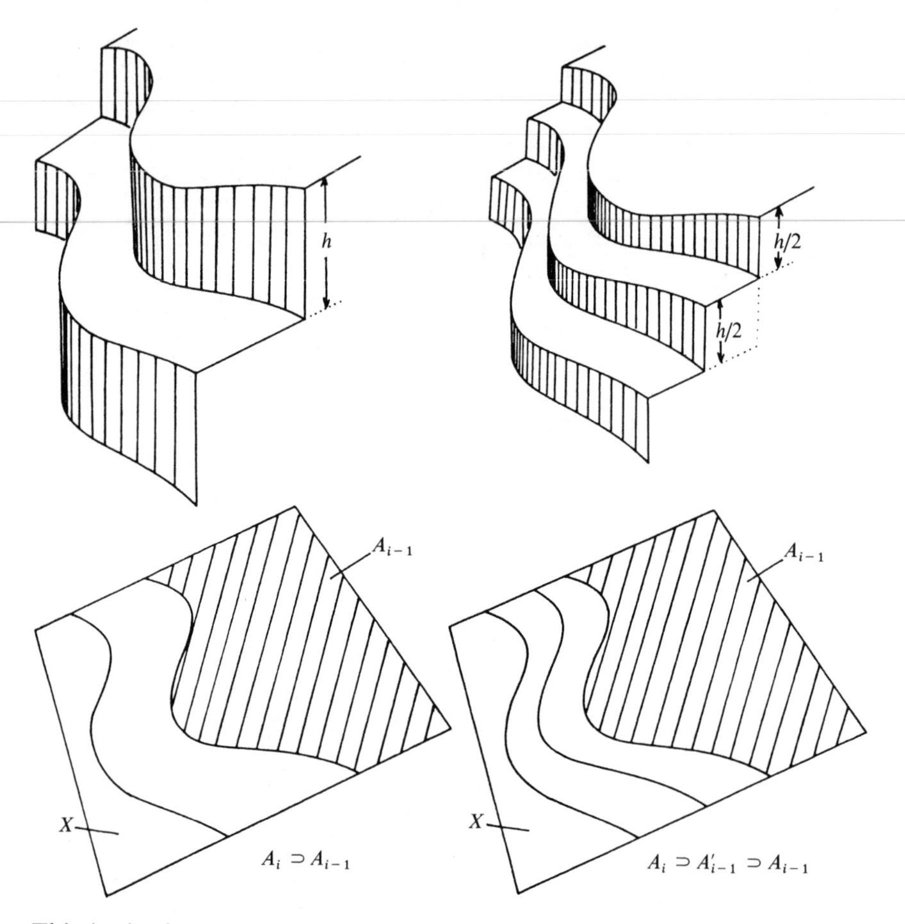

This is the fundamental idea. If we want this procedure to be successful, though, we must make sure that the boundary of A_{i-1} never touches the boundary of A_i, because at a point where they touch the leap would already be higher than its "nominal value" h, and above all it would continue to be greater than h no matter how many times we indent new plateaus:

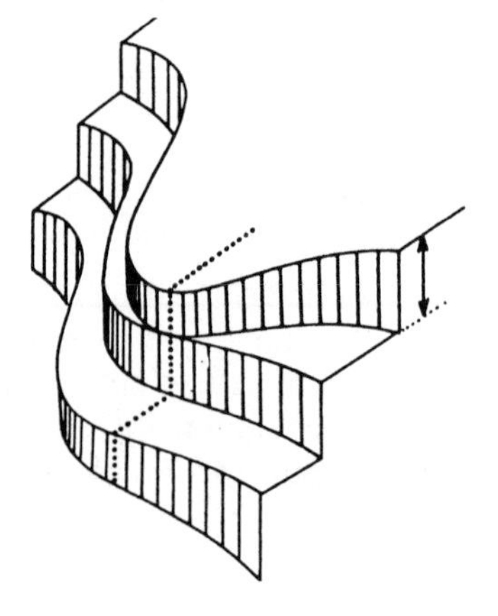

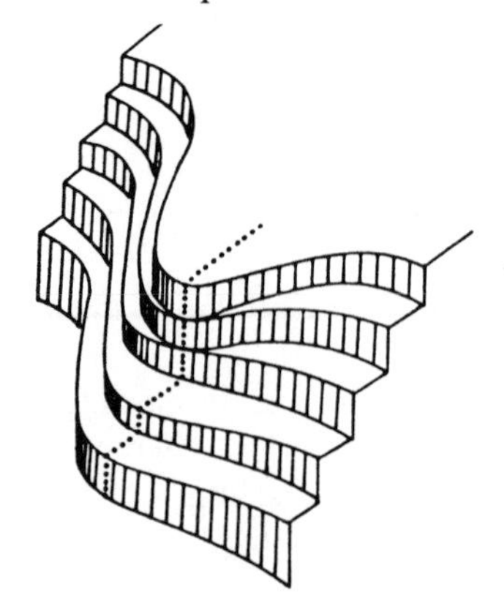

We must thus guarantee that the closure of A_{i-1} is always contained in the interior of A_i, that is $\bar{A}_{i-1} \subset \mathring{A}_i$. When we start the induction, our chain contains only the two sets $A =: A_0 \subset A_1 := X \backslash B$, and the condition is obviously satisfied. Maintaining it along the inductive refining of the chain is exactly where the separation axiom comes into the proof:

Note. *If X is such that any two disjoint closed subsets of X can be separated by open neighborhoods, then for any two subsets M, N with $\bar{M} \subset \mathring{N}$ there is a third subset L "between" them, such that $\bar{M} \subset \mathring{L} \subset \bar{L} \subset \mathring{N}$; just separate M $X \backslash N$ by open neighborhoods U and V and put $L := U$.*

*

This is then the idea of the proof. $\bar{A}_{i-1} \subset \mathring{A}_i$ is an obvious precaution: do we have to take any others? If we did, we'd notice it immediately when trying to carry out our idea—but in fact there are no other obstacles at all, and the proof goes smoothly without any tricks. Let's convince ourselves:

Proof of the Urysohn Lemma. Let A and B be disjoint closed subsets of X. An increasing chain $\mathfrak{A} = (A_0, \ldots, A_r)$ of subsets of X with

$$A = A_0 \subset A_1 \subset \cdots \subset A_r \subset X \backslash B$$

will be called *admissible* if $\bar{A}_{i-1} \subset \mathring{A}_i$ for all i. The function $X \to [0, 1]$ that takes the constant value 1 on A, the value $1 - k/r$ on $A_k \backslash A_{k-1}$ and the value 0 outside A_r will be called the *uniform step function* of the chain $\mathfrak{A}$. The open sets $\mathring{A}_{k+1} \backslash \bar{A}_{k-1}$, $k = 0, \ldots, r$, where $A_{-1} = \varnothing$ and $A_{r+1} = X$, are called the *step domains* of $\mathfrak{A}$, because of their geometrical meaning. Notice that the step domains of an admissible chain cover the whole space, because

$$\bar{A}_k \backslash \bar{A}_{k-1} \subset \mathring{A}_{k+1} \backslash \bar{A}_{k-1}.$$

Notice also that the uniform step function does not fluctuate by more than

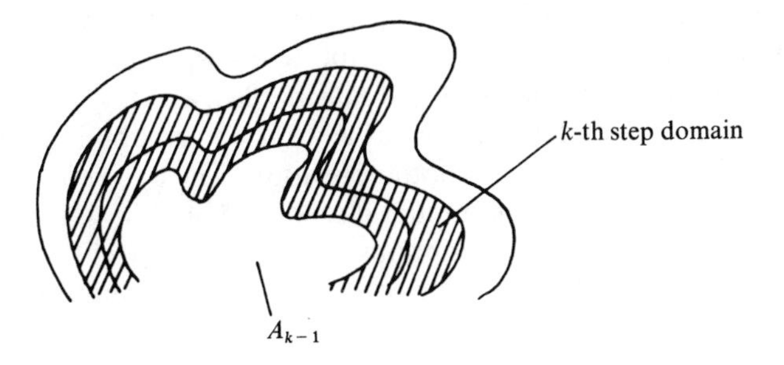

$1/r$ on each step domain. Finally, by a *refinement* of an admissible chain $(A_0, \ldots, A_r)$ we'll mean an admissible chain $(A_0, A'_1, A_1, \ldots, A'_r, A_r)$. As shown in the note above, the separation property of the space guarantees that every admissible chain can be refined.

Now let $\mathfrak{A}_0$ be the admissible chain $(A, X \setminus B)$, and $\mathfrak{A}_{n+1}$ be a refinement of $\mathfrak{A}_n$ for each n. Call f_n the uniform step function of $\mathfrak{A}_n$. Then the following obviously holds: The function sequence $(f_n)_{n>1}$ is pointwise monotonically increasing and bounded by the value 1; in particular it is pointwise convergent and the limit function $f := \lim_{n\to\infty} f_n: X \to [0, 1]$ has the desired property $f|A = 1$ and $f|B = 0$, because each f_n does. There remains to prove the continuity of f. To do this, notice that $|f(x) - f_n(x)|$ is always $\leq \sum_{k=n+1}^{\infty} 1/2^k = 1/2^n$, and that f_n does not fluctuate by more than $1/2^n$ on every step domain of $\mathfrak{A}_n$. Thus f itself fluctuates by no more than $1/2^{n-1}$ on a step domain, and this implies continuity: If $\varepsilon > 0$ and $x \in X$, choose an n such that $1/2^{n-1} < \varepsilon$, and the whole step domain of $\mathfrak{A}_n$ containing x (an open neighborhood of x!) will be mapped into $(f(x) - \varepsilon, f(x) + \varepsilon)$, and f is continuous, qed. □

§3. The Tietze Extension Lemma

The Urysohn lemma may appear at first glance to be a bit too specific, but it can do more than providing functions that are 0 and 1 at certain places. In particular it has the following important consequence and generalization:

Tietze extension lemma. *Suppose that in a topological space X any two disjoint closed sets can be separated by open neighborhoods. Then every continuous function $f: A \to [a, b]$ defined on a closed set A can be extended to a continuous function $F: X \to [a, b]$.*

Proof. Just for this proof let's introduce the following terminology: If $\varphi: A \to \mathbb{R}$ is a bounded continuous function and $s := \sup_{a\in A} |\varphi(a)|$, a continuous function $\Phi: X \to [-s/3, s/3]$ is called a "$\frac{1}{3}$-close approximate extension" of φ if $|\varphi(a) - \Phi(a)| \leq 2s/3$ for all $a \in A$. Such a Φ is thus not a real solution to the extension problem for φ, but just a coarse approximation. The existence of such $\frac{1}{3}$-close extensions is what we can obtain directly from the Urysohn lemma with a one-shot application: The two sets $\varphi^{-1}([s/3, s])$ and $\varphi^{-1}([-s, -s/3])$ are disjoint and closed in A, and in X too, since A is itself closed. There is then a continuous function $X \to [0, 1]$ that takes the values 1 and 0 on these sets, and from it we obtain a continuous map

$$\Phi: X \to [-s/3, s/3]$$

taking the values $s/3$ and $-s/3$ on the same sets; such a function is obviously a $\frac{1}{3}$-close extension of φ.

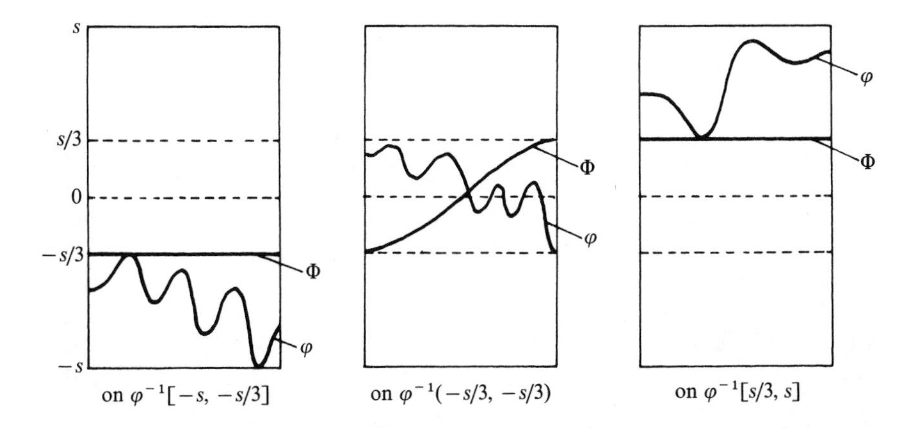

Now to the construction of F. Without loss of generality, we can assume $[a, b] = [-1, 1]$. We first choose a $\frac{1}{3}$-close approximate extension F_1 of f, then a $\frac{1}{3}$-close approximate extension of the "error" $f - F_1|A$; the sum $F_1 + F_2$ is then a better approximate extension. We continue to proceed inductively: Let F_{n+1} be a $\frac{1}{3}$-close approximate extension of

$$f - (F_1 + \cdots + F_n)|A.$$

Then we obviously have:

$$|f(a) - (F_1(a) + \cdots + F_n(a))| \leq (\tfrac{2}{3})^n \quad \text{for all } a \in A, \quad \text{and}$$

$$|F_{n+1}(x)| \leq \tfrac{1}{3}(\tfrac{2}{3})^n \quad \text{for all } x \in X.$$

Thus $\sum_{n=1}^{\infty} F_n$ converges uniformly towards a continuous extension $F: X \to [-1, 1]$ of f, qed. □

The Tietze extension lemma, too, goes further than the basic version just proved.

Corollary 1. *The Tietze extension lemma evidently remains valid if instead of considering the interval* $[a, b]$ *in* $\mathbb{R}$ *one considers the parallelepiped*

$$[a_1, b_1] \times \cdots \times [a_n, b_n]$$

in $\mathbb{R}^n$ *as counterdomain (apply the basic version to each component function), and consequently to every counterdomain homeomorphic to such a parallelepiped, for instance, the closed ball* D^n:

$$f: A \to \mathbb{R}^n \quad \textit{with } |f(a)| \leq r \textit{ can be continuously extended to}$$

$$F: X \to \mathbb{R}^n \quad \textit{with } |F(x)| \leq r.$$

Corollary 2. *The Tietze extension lemma holds also with* $\mathbb{R}$ *(hence also* $\mathbb{R}^n$*) as counterdomain, instead of* $[a, b]$.

PROOF (from [7], p. 17). First extend $\varphi := \arctan f: A \to (-\pi/2, \pi/2)$ to $\Phi: X \to [-\pi/2, \pi/2]$. Now of course one cannot take $\tan \Phi$ immediately,

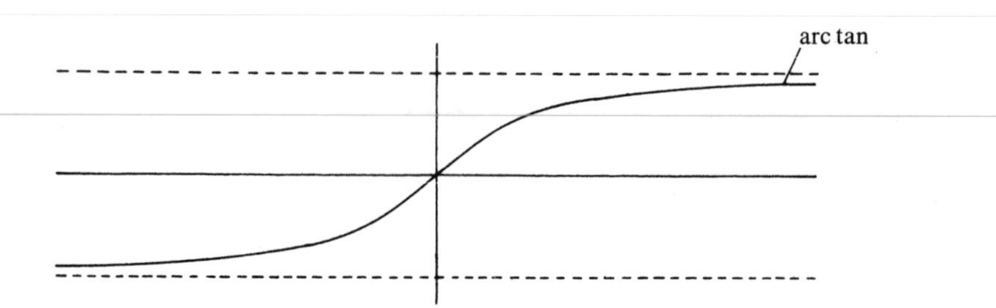

since Φ may actually take the values $\pm \pi/2$. But where? It has to be only on a closed set B disjoint from A. Then if $\lambda: X \to [0, 1]$ is continuous with $\lambda | A \equiv 1$ and $\lambda | B = 0$ (Urysohn lemma), the product $\lambda\Phi: X \to (-\pi/2, \pi/2)$ is a continuous extension of $\arctan f$ not taking the values $\pm \pi/2$, and $\tan \lambda\Phi =: F$ is a continuous extension of f, qed. □

§4. Partitions of Unity and Vector Bundle Sections

Definition (Partition of Unity). Let X be a topological space. A family $\{\tau_\lambda\}_{\lambda \in \Lambda}$ of continuous functions $\tau_\lambda: X \to [0, 1]$ is called a *partition of unity* if: (1) it is "locally finite", in the sense that every point $x \in X$ has a neighborhood in which the τ_λ vanish for all but a finite number of λ: and (2) for every $x \in X$, we have

$$\sum_{\lambda \in \Lambda} \tau_\lambda(x) = 1.$$

The partition of unity is said to be *subordinate* to a given open cover $\mathfrak{U}$ of X if for every λ the support of τ_λ, i.e. the closure

$$\operatorname{Supp} \tau_\lambda := \overline{\{x \in X \mid \tau_\lambda(x) \neq 0\}}$$

is entirely contained in one of the sets of the covering.

Partitions of unity shall occupy us for the rest of the chapter. How they can be obtained will be discussed in the next sections; right now we will just talk about what can be done with them when they are already available. To this end I will first go into the construction of sections in vector bundles, because this illustrates a principle that is typical of the application of partitions of unity in a number of individual examples. For starts, let me offer a very short

EXCURSUS ON VECTOR BUNDLES AND THEIR SECTIONS

Definition (Vector Bundle). The data of an *n-dimensional real vector bundle* over a topological space X consist of three parts:

(i) a topological space E (called "total space");
(ii) a continuous surjective map $\pi: E \to X$ ("projection"); and
(iii) a real vector space structure on each "fiber" $E_x := \pi^{-1}(x)$.

For these data to make up an n-dimensional vector bundle over X, they only have to satisfy one axiom:

Axiom (Local Triviality). For every point in X there is a "bundle chart", or simply "chart", (h, U), i.e. an open neighborhood U and a homeomorphism

$$\pi^{-1}(U) \xrightarrow[\cong]{h} U \times \mathbb{R}^n,$$

whose restriction to F_x, for every $x \in U$, is a linear isomorphism onto $\{x\} \times \mathbb{R}^n$.

Definition (Sections in Vector Bundles). A continuous map $\sigma: X \to E$ taking each point to an element in its fiber (i.e. such that $\pi \circ \sigma = \mathrm{Id}_X$) is called a *section* of E. In particular, for every vector bundle the map $\sigma: X \to E$ taking each x to the origin in E_x is a section ("zero section").

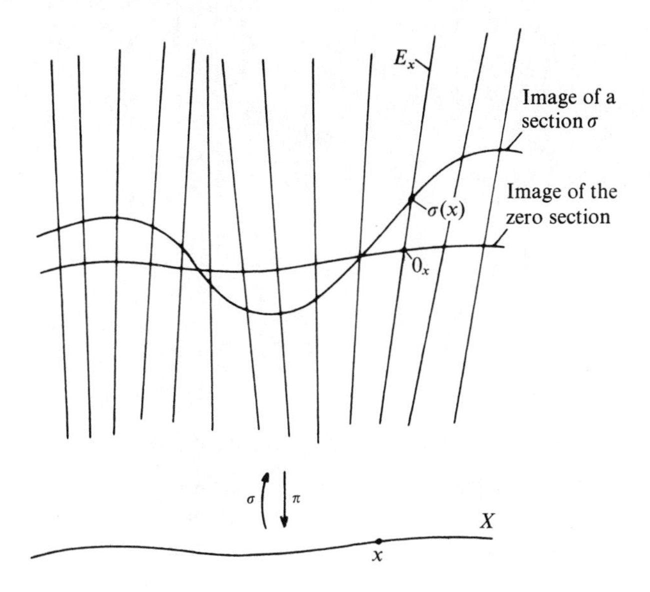

Asked about the "most important" examples of n-dimensional real vector bundles, I would answer without hesitation: the tangent bundles $TM \xrightarrow{\pi} M$ of n-dimensional differentiable manifolds M. The sections of tangent bundles are exactly the tangent vector fields on M. Also many other objects in analysis and geometry, for instance alternating differential forms, Riemann metrices and various other "co- and contravariant tensor fields", as they are known in a somewhat obsolescent terminology, are sections of bundles derived from the tangent bundle.

But it is not only over manifolds that vector bundles and their sections must be considered, but also over more general topological spaces, not least because the vector bundles over X lead to a functor K ("K-theory") from the topological category into the category of rings which is now indispensable (although the way this comes about cannot be foreshadowed by our remarks here). I will not even start to speak about the role of K-theory, just give a reference: M. Atiyah, *K-Theory*, New York–Amsterdam 1967—in this book, by the way, you will find the Tietze extension theorem and partitions of unity being used as instruments right from the start (§1.4). End of the very short (as promised) excursus on vector bundles and their sections.

Let then $E \xrightarrow{\pi} X$ be a vector bundle over X, and let's suppose we want to construct a section $f : X \to E$—with certain properties, of course, otherwise we could simply take the zero section. Let's suppose also that the problem is locally solvable, using for example charts. Then we can find an open cover $\mathfrak{U}$ of X such that for every set U in the cover there is a "local solution" to our problem, i.e. a section $U \to \pi^{-1}(U)$ with the desired properties.

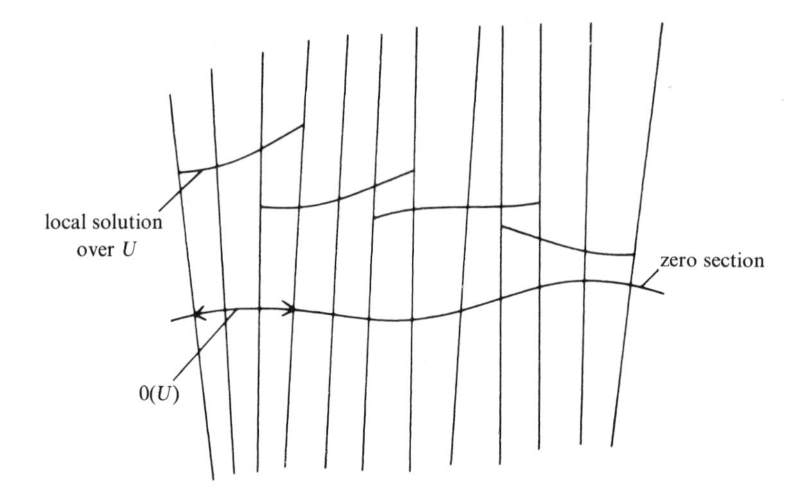

Now comes in the partition of unity. We choose (if possible, see §5) a partition of unity $\{\tau_\lambda\}_{\lambda\in\Lambda}$ subordinate to $\mathfrak{U}$, then for every λ we choose a set U_λ in $\mathfrak{U}$ that contains the support of τ_λ, then a local solution

$$f_\lambda : U_\lambda \to \pi^{-1}(U_\lambda).$$

Then it is clear how $\tau_\lambda f_\lambda$, which is initially defined only on U_λ, can be considered as a continuous section on the whole of X: just extend it with the value 0 outside U_λ, and because of the local finiteness of the partition of unity we get by the formula

$$f := \sum_{\lambda \in \Lambda} \tau_\lambda f_\lambda$$

a global continuous section $f : X \to E$ which so to speak interpolates as best it can the local solutions: If at a point $x \in X$ all the f_λ defined there have the same value $f_\lambda(x)$, so does f because $\sum_{\lambda \in \Lambda} \tau_\lambda \equiv 1$; if they have different values to choose from, however, f averages them off with "weights" $\tau_\lambda(x)$, $\lambda \in \Lambda$. The question is now, under what circumstances are the desired properties of the local solutions f_λ transferred to the global section f by this procedure?

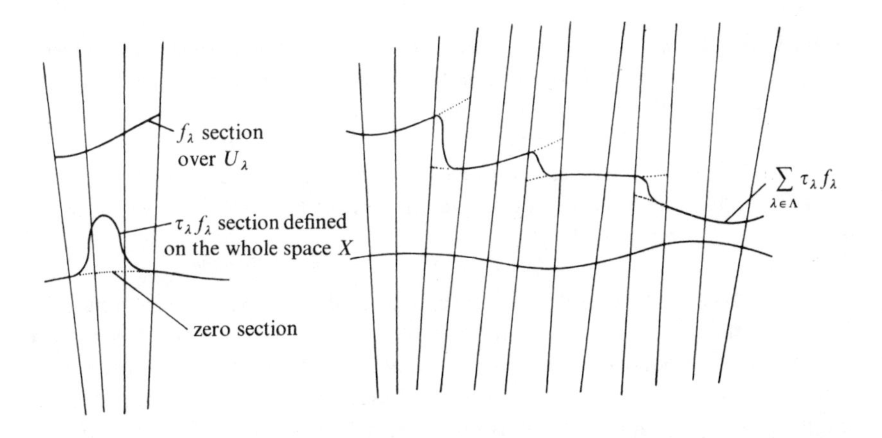

For some applications this can only be achieved by skilfully selecting the f_λ and τ_λ; just take the following oversimplified example to see why this may be the case:

$x = \mathbb{R}$, $E = \mathbb{R} \times \mathbb{R}$, property: "monotonically increasing"

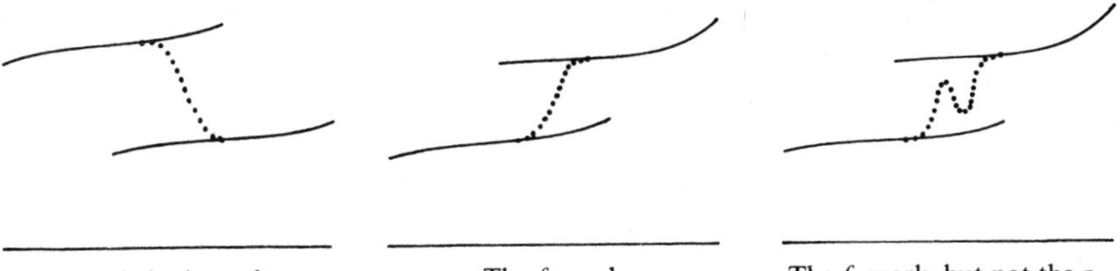

But we won't discuss such cases here, just the numerous other cases in which the desired property is automatically transferred from the f_λ to f, the so-called "convex properties". The nomenclature is because $\tau_\lambda(x) \in [0, 1]$ and $\sum \tau_\lambda = 1$ means that for every x the image $f(x)$ is in the convex hull of a finite number of $f_\lambda(x)$:

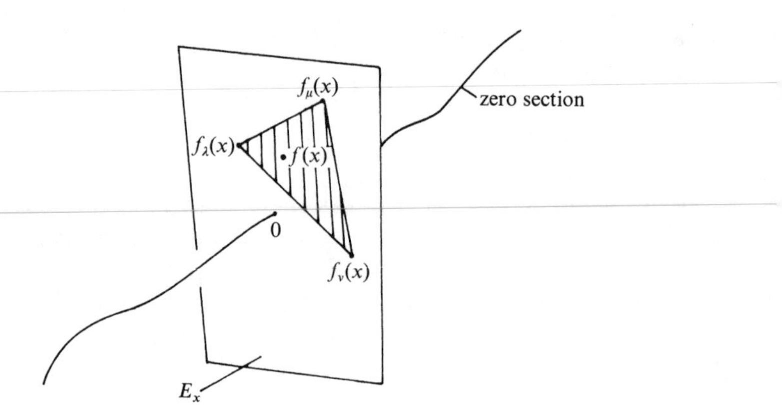

Note. *Let $\pi: E \to X$ be a vector bundle over a topological space X for which there is a partition of unity subordinate to every open cover. Furthermore, let $\Omega \subset E$ be a fiberwise convex set, i.e. suppose every $\Omega_x := \Omega \cap E_x$ is convex, and assume finally that there are local sections of E sitting in Ω, i.e. that every point in X has an open neighborhood U and a section $U \to \pi^{-1}(U)$ whose image lies in Ω. Then there is also a global section $f: X \to E$ whose image lies in Ω.*

When this sort of reasoning is applied, one generally says just: "There are local sections with such and such property, and since the property is convex, we obtain using partitions of unity a global section with the same property." This is an excellent shorthand expression, that avoids the trouble of going into the nitty-gritty of detailed notation. We'll now consider a few examples of such convex properties. Notice also that several convex properties taken together form again a convex property (the intersection of convex sets is convex).

(1) The property of coinciding on a set $A \subset X$ with a section $f_0: A \to \pi^{-1}(A)$ already given on it is convex: For $a \in A, \Omega_a = \{f_0(a)\}$, for $x \notin A, \Omega_x$ is the whole fiber E. Thus if we know that f can be locally extended, for instance using the Tietze extension lemma, we get a global extension using a partition of unity.
(2) Let X be a differentiable manifold, $E := TX$. The property of vector fields on X (i.e. sections in TX) of being tangential to one or more given submanifolds is convex.

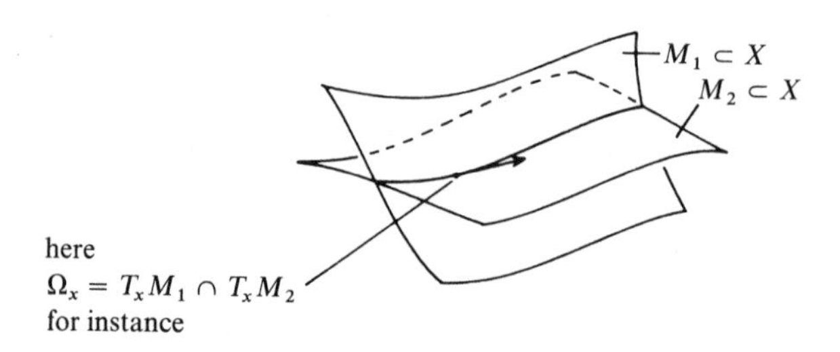

(3) In connection with (1) the following observation is often useful: Let M be a differentiable manifold, $X := M \times [0, 1]$, $E := TX$. The property of a vector field on X having as its component in the direction $[0, 1]$ the standard unit vector $\partial/\partial t$ is convex. The flow of such a vector field takes $M \times 0$ onto $M \times t$ for every time t; in this way one builds in differential topology "diffeotopies", i.e. differentiable homotopies

$$H: M \times [0, 1] \to M$$

for which every particular $H_t: M \to M$ is a diffeomorphism and $H_0 = \mathrm{Id}_M$. Of course the point is not finding any diffeotopy, but rather one that achieves some particular end, for instance, carrying a prescribed "isotopy" $h: N \times [0, 1] \to M$ (each h_t an embedding)

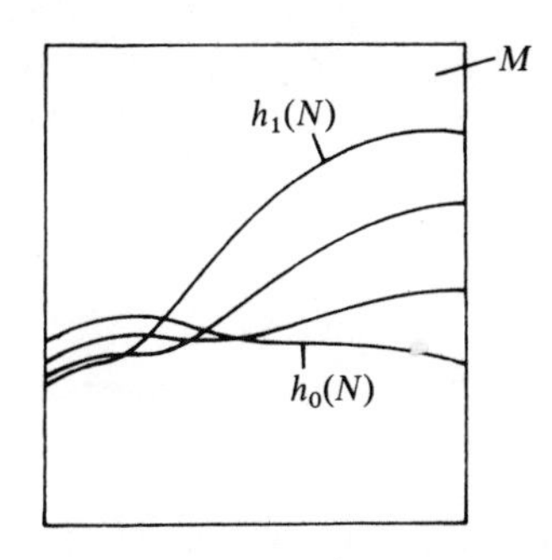

with itself: $H_t \circ h_0 = h_t$. This problem leads to the question of finding a vector field on $M \times [0, 1]$ which lies "over " $\partial/\partial t$ (as in (3) above) and which is already prescribed on the submanifold $\bigcup_{t \in [0,1]} h_t(N) \times t$ given by the isotopy (as in (1)).

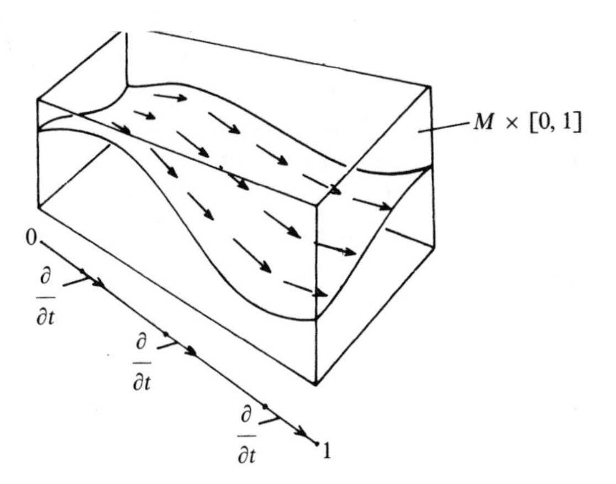

For details see for instance [3], §9.

All in all, a differential topologist would be entirely lost without partitions of unity, because the many diffeomorphisms used in differential topology are almost always obtained by integration of vector fields, and the vector fields are almost always secured by local construction (analysis) and partitions of unity (topology).

(4) Let E be a vector bundle over a topological space X. The property of sections in the vector bundle $(E \otimes E)^*$ (bilinear forms on the fibers of E) of being symmetric and positive definite is convex. Thus we get "Riemannian metrics" on vector bundles, in particular on tangential bundles TM ("Riemannian manifolds").

(5) Let E be a vector bundle over X with a Riemannian metric on each fiber. Let $\varepsilon > 0$ and $\sigma: X \to E$ a prescribed section. The property of sections in E of remaining inside the "ε-tube" around σ is convex:

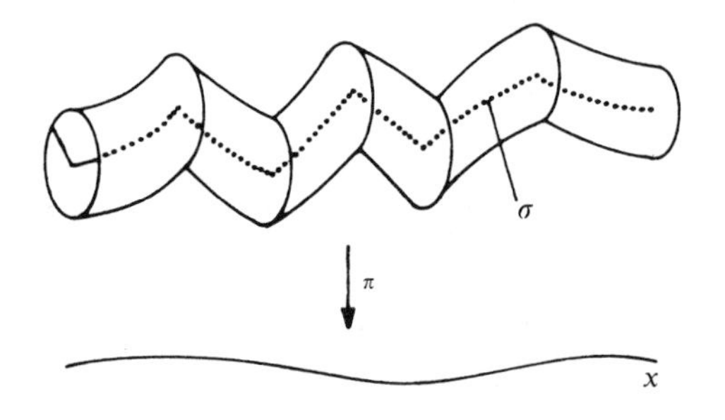

Here again the point is not just finding a section that remains in the tube, since σ is one in the first place, but getting sections with additional, "better" properties than σ, that approximate σ "within ε".

*

Though these examples are entirely typical of the applications of partitions of unity, I still must straighten the picture given by them with some additional strokes. First, you shouldn't get the impression we're always in the "local situation" just because we have a chart before us. This is sometimes the case (as in (4)), but in general the "local theory" says only that every point possesses a (possibly very small) neighborhood on which there is a local solution. In this case partitions of unity are already indispensable in the harmless case that X is itself an open subset of $\mathbb{R}^k$ and E is simply $X \times \mathbb{R}^n$. Singularity theory, for example, often leads to this situation.

Secondly: The construction of global objects starting from local data is certainly the main objective of partitions of unity. But they can also be used to break up existing global objects into local ones and thus make computations practicable. If for instance $M \subset \mathbb{R}^n$ is, say, a compact k-dimensional submanifold, and $f: M \to \mathbb{R}$ is, say, a continuous function on M, the integral $\int_M f\, dV$ can be defined and studied by choosing a finite partition of unity subordinate to a finite atlas. This reduces each individual $\int_M \tau_\lambda f\, dV$ (using charts) to the usual multiple integral in $\mathbb{R}^n$

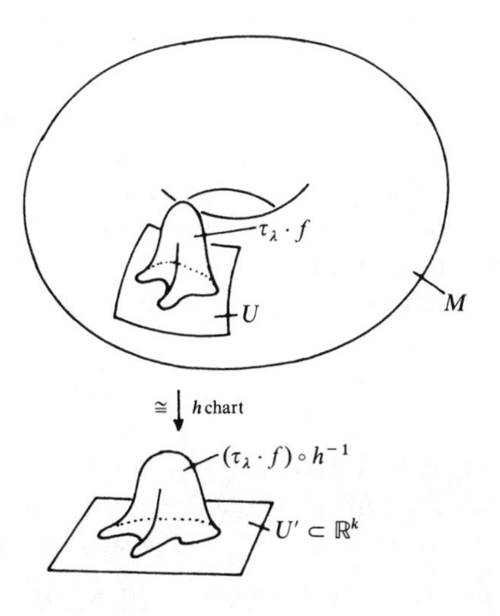

(the actual formula is $\int_M \tau_\lambda f\, dV = \int_{\mathbb{R}^k} (\tau_\lambda f)h^{-1} \cdot \sqrt{g}\, dx_1 \cdots dx_n$, where g is the determinant of the fundamental form of the metric (g_{ij})). The final step is to add everything up via

$$\int_M f\, dV = \sum_{\lambda \in \Lambda} \int_M \tau_\lambda f\, dV.$$

Thirdly and finally, let us mention that partitions of unity don't exist just for the sake of functions and sections in vector bundles, but fulfil many other subtler ends. See for instance A. Dold, "Partitions of unity in the theory of fibrations", 1963 [6].

§5. Paracompactness

Only with some hesitation do I introduce yet another topological concept: paracompactness. There are so many such concepts! An A is called B if for every C there is a D such that E holds—this is quite boring in the beginning, and remains so until we can see some *sense* behind it, until we can see the spirit behind the letter. To define a first uninteresting property, then a second uninteresting property, just to say that the first uninteresting property implies the second uninteresting property, but there is an uninteresting example that satisfies the second uninteresting property but not the first—good heavens! Never was a meaningful concept introduced in mathematics in a random or just playful way: the sense is there first, and the ends *create* the means.

Now I know as well as the next person that in academic teaching it is entirely unavoidable to put students off with a "later", and that often; formal and technical knowledge must reach a certain level before we can start talking about the meaning of things honestly, that is, without substituting simple but false motives for the true complex ones. But "as formal as necessary" already means very formal in mathematics; and things shouldn't be made any more formal than that. If someone is expected too often to find preparations towards unknown objectives interesting, he eventually loses interest in knowing what the objectives are in the first place; and I'm afraid many a student graduates from our universities without having seen the central fire of mathematics glow anywhere, and, what is worse: without even believing the existence of such a central fire.—But I'm straying too far from my subject.

Definition (Paracompact). A Hausdorff space X is called paracompact if every open cover has a locally finite subcover, i.e. if for every open cover $\mathfrak{U}$ of X there is an open cover $\mathfrak{V} = \{V_\lambda\}_{\lambda \in \Lambda}$ of X such that the following holds:

(1) $\mathfrak{V}$ is locally finite, i.e. every $x \in X$ has a neighborhood that intersects V_λ for only finitely many λ; and
(2) $\mathfrak{V}$ is a refinement of $\mathfrak{U}$, i.e. every V_λ is contained in a set of $\mathfrak{U}$.

This is, so to speak, the "boring state" of the concept. It immediately acquires interest, though, because of the

Theorem. *A Hausdorff space is paracompact if and only if it has the nice property that every open cover admits a partition of unity subordinate to it.*

One direction of the proof is trivial: If $\{\tau_\lambda\}_{\lambda \in \Lambda}$ is a partition of unity subordinate to $\mathfrak{U}$, then the $V_\lambda := \{x \in X \mid \tau_\lambda(x) \neq 0\}$ form a locally finite refinement of $\mathfrak{U}$. The proof of the converse, i.e. every paracompact space has the "partition-of-unity property", will close the paragraph and the chapter. But first I will go into an obvious question, namely: What is the theorem for? Why not *define* the concept by using the assertion of the theorem, if the whole point is to have partitions of unity? Now, exactly because the partition-of-unity property is so wonderful one would like to be able to recognize as many spaces as possible as having it, and this is better done with help of the theorem than directly. For instance, every compact Hausdorff space is trivially paracompact, but is it immediately clear that it satisfies the partition-of-unity property? No, only using the theorem. The following information, given without proof, will show you that paracompactness is a widespread, "common" property.

Remark. If a Hausdorff space is locally compact, i.e. if every neighborhood contains a compact neighborhood, and if moreover it is the union of countably many compact subspaces (which due to local compactness is true for instance for second countable spaces), then it is paracompact.

Corollary. *Manifolds, in particular* $\mathbb{R}^n$, *are paracompact.*

Remark. The product of a paracompact space and a compact Hausdorff space is paracompact.

Theorem (Stone). *Every metrizable space is paracompact!*

In particular all subspaces of metrizable spaces are paracompact, because they are again metrizable. But this is remarkable, because in general paracompactness is inherited by closed subsets (same reason as for compact spaces, see Chapter I), but not by arbitrary subspaces.

Theorem (Miyazaki). *Every* CW*-complex is paracompact.*

For the proof of this last theorem see the reference in [5]; for the others see for instance [16], Chapter I, §§8.5 and 8.7. The proofs are not at all difficult—to read.

But now to the promised proof that in a paracompact space every open cover has a subordinated partition of unity. The proof has two parts:

(1) **Lemma**. *In every paracompact space the Urysohn lemma is applicable, i.e. every two closed disjoint subsets can be separated by open neighborhoods.*
(2) *Construction of partitions of unity using the Urysohn lemma.*

For (1). Let A and B be disjoint closed subspaces of the paracompact space X. For every two points $a \in A$, $b \in B$ we choose separating open neighborhoods $U(a, b)$ and $V(a, b)$. Now we keep a fixed and try to separate a and B by open neighborhoods $U(a)$ and $V(a)$. To this end we choose a locally

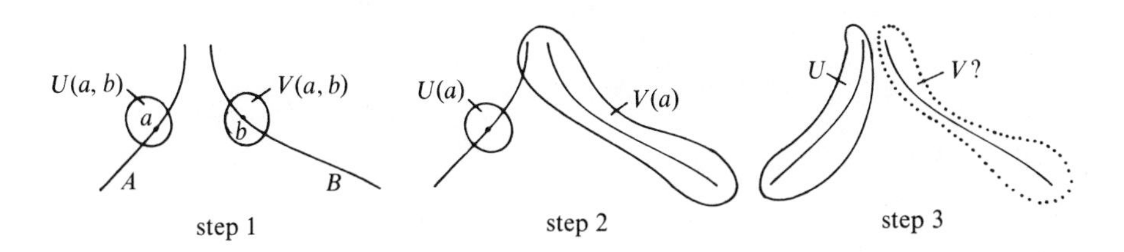

finite refinement of the open cover given by $\{V(a, b)\}_{b \in B}$ and $X \setminus B$, and define $V(a)$ as the union of all sets of this refinement that are contained in some $V(a, b)$, $b \in B$. Because of local finiteness there is now an open neighborhood of a that intersects only finitely many such refinement sets, and if they lie in $V(a, b_1) \cup \cdots \cup V(a, b_r)$, we just have to take the intersection of this neighborhood of a with $U(a, b_1) \cap \cdots \cap U(a, b_r)$ to obtain a neighborhood $U(a)$ of a disjoint from $V(a)$.

Now instead of keeping a fixed we choose as above a locally finite refinement of the open cover given by $\{U(a)\}_{a \in A}$ and $X \setminus A$, and define U as the union of all the sets of this refinement that lie in $U(a)$ for $a \in A$. All we need now is to find for each $b \in B$ an open neighborhood not intersecting U: the union of those is the desired V. In any case b has an open neighborhood that intersects only finitely many of those refinement sets whose union we defined to be U. Suppose those finitely many sets are contained in

$$U(a_1) \cup \cdots \cup U(a_s).$$

Then the intersection of this neighborhood of b and

$$V(a_1) \cap \cdots \cap V(a_s)$$

is the desired neighborhood disjoint from U, qed for (1).

For (2). Now let $\mathfrak{U} = \{U_\lambda\}_{\lambda \in \Lambda}$ be an open cover of X, which can be taken without loss of generality to be locally finite, by (1). To find a partition of unity subordinate to it we will first "shrink" the U_λ a bit, i.e. we will find an open cover $\{V_\lambda\}_{\lambda \in \Lambda}$ with $\overline{V}_\lambda \subset U_\lambda$. For suppose this is possible: Then we choose $\sigma_\lambda: X \to [0, 1]$ with $\sigma_\lambda | \overline{V}_\lambda \equiv 1$ and $\sigma_\lambda | X \backslash U_\lambda \equiv 0$ by the Urysohn lemma, obtaining a locally finite family $\{\sigma_\lambda\}_{\lambda \in \Lambda}$; the sum $\sigma := \sum_{\lambda \in \Lambda} \sigma_\lambda$ is continuous and positive everywhere, and the desired partition of unity would be obtained by setting $\tau_\lambda := \sigma_\lambda / \sigma$, $\lambda \in \Lambda$.

So there remains to prove that $\mathfrak{U}$ has a "shrinking". Choose for every $x \in X$ an open neighborhood Y_x such that $\overline{Y}_x$ is entirely contained in a set of $\mathfrak{U}$. This is possible because by (1) x and $X \backslash U_\lambda$ for $x \in U_\lambda$ can be separated by open neighborhoods. Let $\{W_\alpha\}_{\alpha \in A}$ be a locally finite refinement of $\{Y_x\}_{x \in X}$ and V_λ the union of all the W_α whose closure lies in U_λ. Then of course $\{V_\lambda\}_{\lambda \in \Lambda}$ is again an open cover, and actually $\overline{V}_\lambda \subset U_\lambda$, because: Take $x \in \overline{V}_\lambda$. Then every neighborhood of x intersects at least one W_α whose closure is contained in U_λ. But due to the local finiteness of $\{W_\alpha\}_{\alpha \in A}$ a small enough neighborhood intersects only finitely many of those, say $W_{\alpha_1}, \ldots, W_{\alpha_r}$. But now x must lie in $\overline{W_{\alpha_1} \cup \cdots \cup W_{\alpha_r}}$, otherwise there would be a neighborhood of x that does not intersect any W_α whose closure lies in U_λ. Thus

$$x \in \overline{W_{\alpha_1} \cup \cdots \cup W_{\alpha_r}} = \overline{W}_{\alpha_1} \cup \cdots \cup \overline{W}_{\alpha_r} \subset U_\lambda,$$

qed. □

CHAPTER IX
Covering Spaces

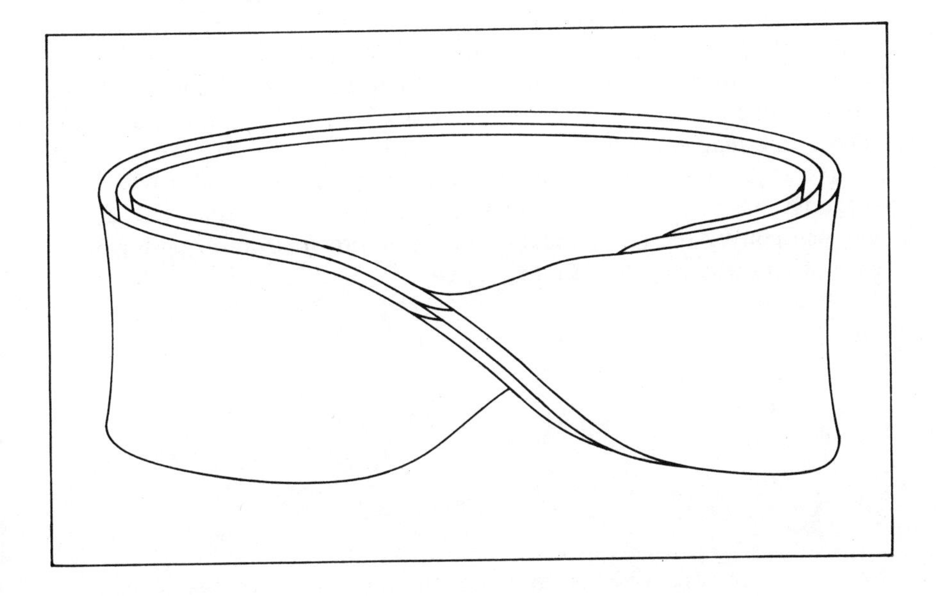

§1. Topological Spaces Over X

A covering space of X is a space Y with a continuous surjective map $\pi\colon Y \to X$ ("covering map"), which locally, around every point of the "base space" X, looks essentially like the canonical map of a disjoint sum of copies of a space onto their original:

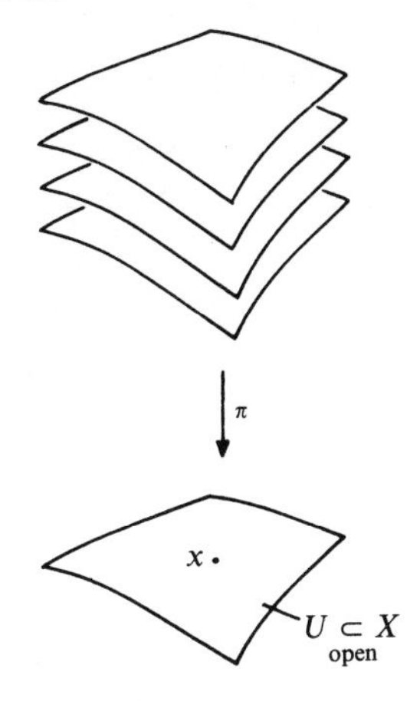

That's it for now—just enough to launch your intuition in the right direction. The exact definition comes in §2.

In the normal intuition associated with a map $f: A \to B$, A is the primary object with which something "happens" under the map: Each point $a \in A$ is "mapped" to an image point $f(a) \in B$. But one could equally well consider the counter-domain B as the primary object and think of the map $f: A \to B$ as a family $\{A_b\}_{b \in B}$ of "fibers" $A_b := f^{-1}(b)$ over B. The two ways of expressing a map or looking at it are, of course, entirely equivalent, and choosing one or the other is just a matter of what purpose you are considering the map for in the first place. For instance, anyone will visualize a curve $\alpha: [0, 1] \to X$ in the first way (knowing for every time t where the image is); whereas for a vector bundle $\pi: E \to X$, for example, the second way will convey the more suitable general impression (knowing for every $x \in X$ what the corresponding fiber E_x is).

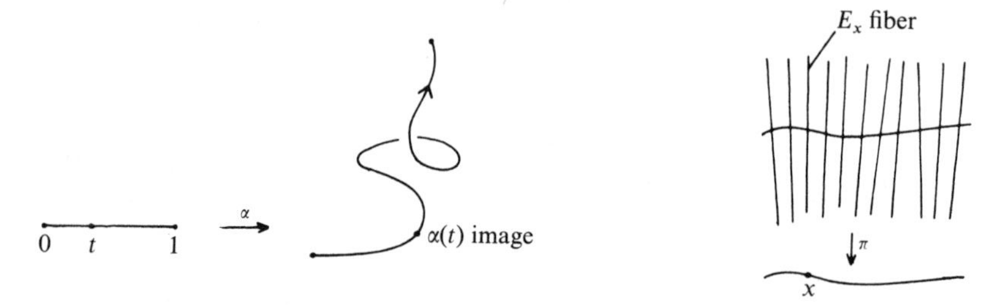

Now for covering spaces it is again with the counterdomain that something "is happening" (it is being covered by something), and for this reason I want to direct your attention to this mode of seeing things by using the following

Terminology ("Over"). Let X be a topological space. By a *topological space over* X we mean a pair (Y, π) consisting of a topological space Y and a continuous surjective map $\pi: Y \to X$. Whenever possible we eliminate π from the notation, i.e. we write, if no confusion will arise,

Y instead of (Y, π)
Y_x instead of $\pi^{-1}(x)$ ("fiber over x")
$Y|U$ instead of $(\pi^{-1}(U), \pi|\pi^{-1}(U))$ ("restriction of Y to $U \subset X$").

Example. The projection on the x-coordinate makes D^2 into a topological space over $[-1, 1]$. The fibers are intervals or (at each end) points:

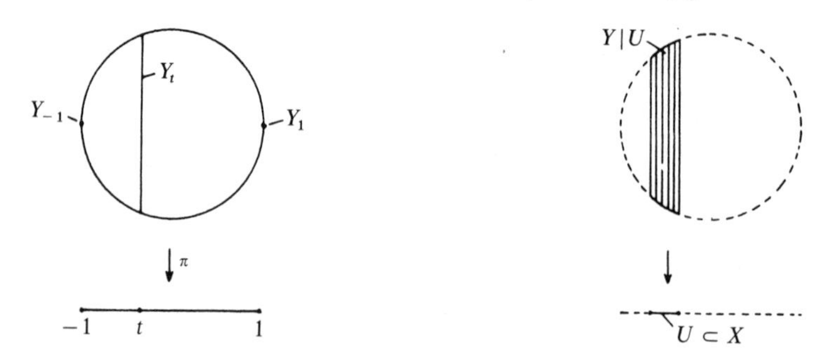

So this is at first just another way of referring to surjective continuous maps. However, the point of view under which they will be considered here will become clear when I indicate when two topological spaces over X are to be seen as equivalent:

Definition. Two topological spaces Y and $\tilde{Y}$ over X will be called homeomorphic over X, or "isomorphic" for short ($Y \cong \tilde{Y}$) if there exists between them a homeomorphism $h: Y \to \tilde{Y}$ "over X", i.e. one for which the diagram

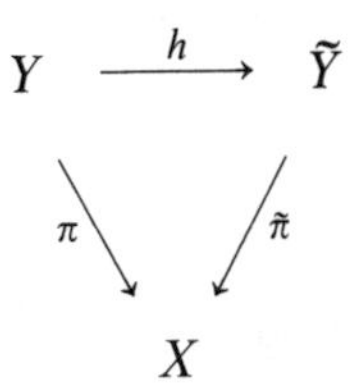

commutes. Notice that then h necessarily maps the fiber Y_x homeomorphically onto the fiber $\tilde{Y}_x$.

Without further restrictive axioms the concept of a topological space over X would still be too general to be seriously considered of any use. A more special class of topological spaces over X, still big but already interesting, is given by the requirement of "local triviality":

Definition (Trivial and Locally Trivial Fibrations). A topological space Y over X is called *trivial* if there is a topological space F such that Y is isomorphic to

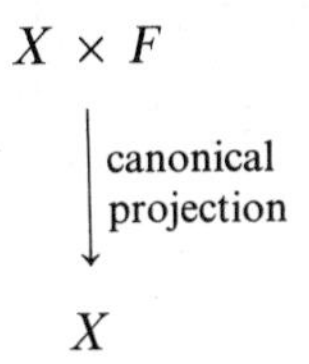

A topological space Y over X is called *locally trivial* or a *locally trivial fibration* if every $x \in X$ has a neighborhood U over which Y is trivial, i.e. such that $Y|U$ is trivial.

If for some neighborhood U of x the restriction $Y|U$ is trivial, then of course it is even true that $Y|U \cong U \times Y_x$. If for a locally trivial fibration all the fibers Y_x are homeomorphic to a fixed space F, Y is called a locally trivial fibration with "typical fiber" F. This is not such a strong restriction as it might seem: the homeomorphism type of the fibers Y_x of a locally trivial fibration is of course locally constant, and consequently globally constant if the base space X is connected.

For locally trivial fibrations Y over X with fiber F there are for instance close ties between the homotopy groups of fiber F, base X and total space Y ("exact homotopy sequence"), so that one can obtain information about the homotopy properties of one of these spaces from the properties of the other two. (In fact local triviality is an unnecessarily strong condition for the exact homotopy sequence to hold; the keyword is "Serre fibrations".)

One word about terminology: fiber *bundles*, which we are not at all going to deal with here, "are" among other things locally trivial fibrations, but a fiber bundle is not merely a locally trivial fibration with special properties: Besides additional axioms this notion comprehends also additional data. For instance in vector bundles we require a vector space structure on each fiber (additional data), and one must be able to choose the local trivializations to be linearly isomorphic on the fibers (additional axiom).

§2. The Concept of a Covering Space

Definition (Covering Space). A locally trivial fibration is called a *covering space* if all its fibers are discrete.

A surjective continuous map $\pi: Y \to X$ is thus a covering space if for every $x \in X$ there exist an open neighborhood U of x and a discrete space Λ such that $Y|U$ and $U \times \Lambda$ are homeomorphic over U.

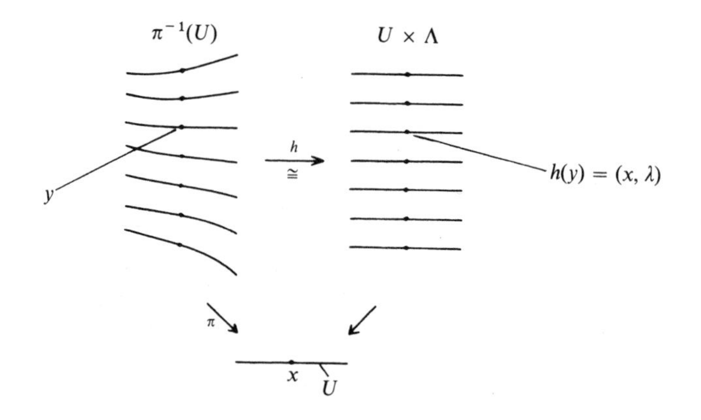

Λ can of course be chosen to be the fiber itself, as is always the case for locally trivial fibrations.

The cardinality $\# Y_x$ of a fiber over x is called the *multiplicity* of the covering at point x. The multiplicity is obviously locally constant and thus for X connected also globally constant. If the multiplicity is constant and equal to n, we talk about an *n-fold covering*.

A covering map $\pi: Y \to X$ is always *locally homeomorphic*, i.e. every $y \in Y$ has an open neighborhood V such that $\pi(V)$ is open in X and π defines a homeomorphism $V \xrightarrow{\cong} \pi(V)$. In $U \times \Lambda$, of course, every $U \times \{\lambda\}$ is open, since Λ is discrete, and the canonical projection $U \times \{\lambda\} \to U$ is clearly a homeomorphism. Thus in the situation depicted above $V := h^{-1}(U \times \lambda)$ is open in $\pi^{-1}(U)$ and hence in Y, and π maps V homeomorphically onto U.

*

This concept of covering space is not the only one in use, only the simplest. In function theory, particularly, one has reason to consider more general "coverings", in particular "branched coverings" like for instance $\mathbb{C} \to \mathbb{C}$, $z \mapsto z^2$:

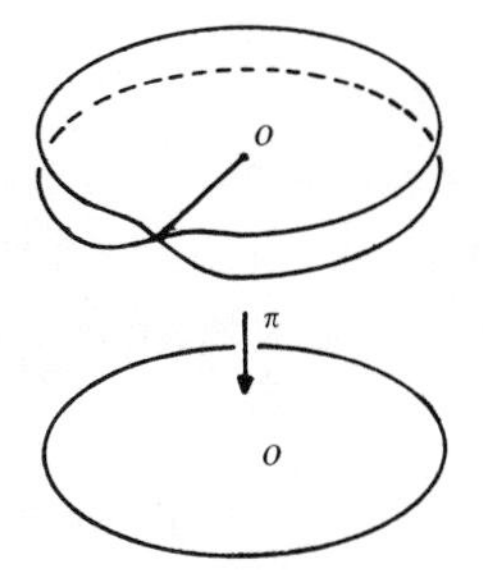

Then local triviality is no longer required, only continuity, openness (images of open sets are open) and discreteness of fibers. (See, for instance, [9], p. 20ff.) The points y at which such a map is not locally homeomorphic are called branching points; in the example $z \mapsto z^2$ the only branching point is 0.

But even when such a "generalized" covering (i.e. a continuous, open, discrete map) in unbranched (that is, everywhere locally homeomorphic) and surjective, it does not have to be a covering space in our sense. It is easy to create examples of this by cutting away an appropriate closed piece from a "good" covering:

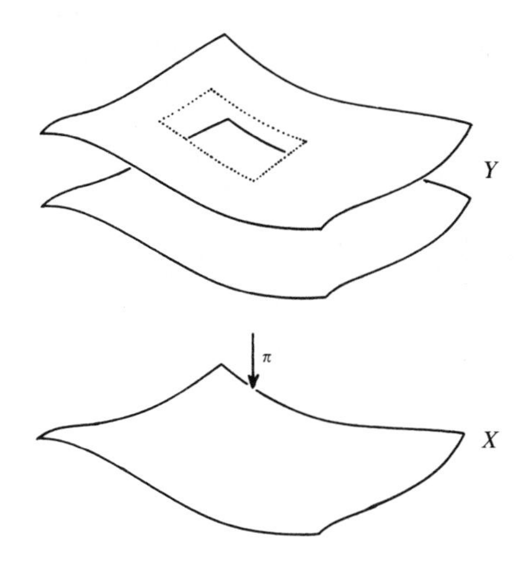

The simplest example is when $Y = \mathbb{R} \times 0 \cup \{x \in \mathbb{R} | x > 0\} \times 1$ with the canonical projection $Y \to \mathbb{R}$: Y is not locally trivial at 0—the number of leaves is 1 to the left and 2 to the right.

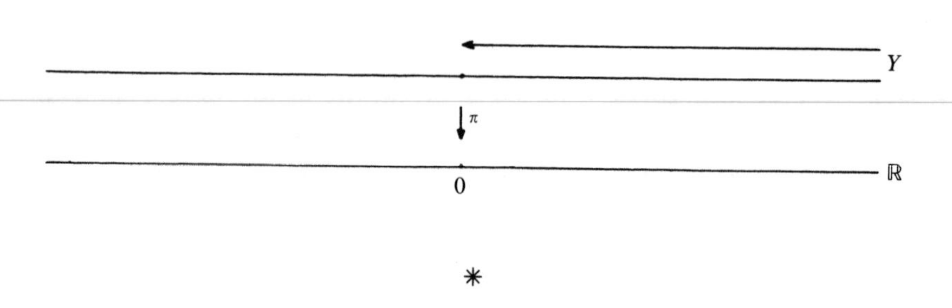

*

To conclude, a couple of very simple but non-trivial (in the technical sense) examples:

(1) For every natural number $n \geq 2$ the map $z \mapsto z^n$ gives an n-fold non-trivial covering $\mathbb{C}\backslash 0 \to \mathbb{C}\backslash 0$ (or $S^1 \to S^1$):

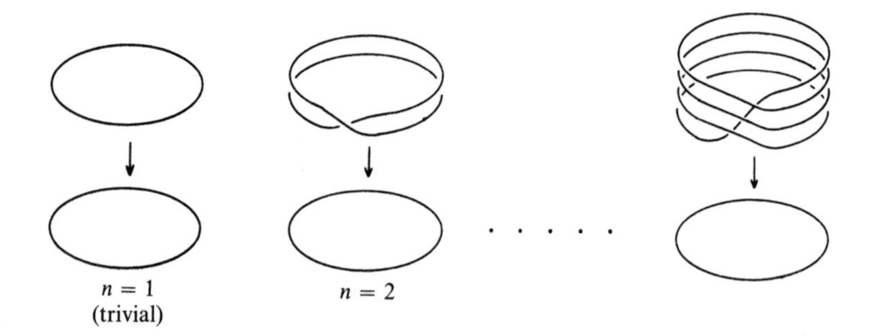

(2) The map $\mathbb{R} \to S^1$, $x \mapsto e^{ix}$, defines a covering with a countably infinite number of leaves.

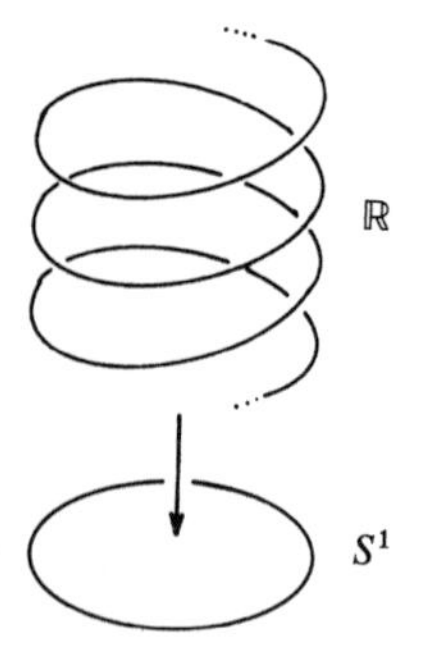

(3) The canonical projection $S^n \to \mathbb{RP}^n$ is a 2-fold covering.

So much for the *notion* itself. The uses of covering spaces I will explain in §8; the major portion of this chapter will be occupied with the classification of covering spaces. Yes, because it is possible to get a sort of complete survey of all covering spaces of a topological space X, and this is what is meant by "covering space theory"—rather, a mini-theory, but a useful one. Now in this theory there is an omnipresent technical tool, with which everything is done, constructed and proved: the lifting of paths. This is what we're going to talk about now.

§3. Path Lifting

Definition (Path Lifting). Let $\pi: Y \to X$ be a covering map and $\alpha: [a, b] \to X$ a continuous map (a "path"). A path $\tilde{\alpha}: [a, b] \to Y$ is called a lifting of α starting at y_0 if $\tilde{\alpha}(a) = y_0$ and $\pi \circ \tilde{\alpha} = \alpha$.

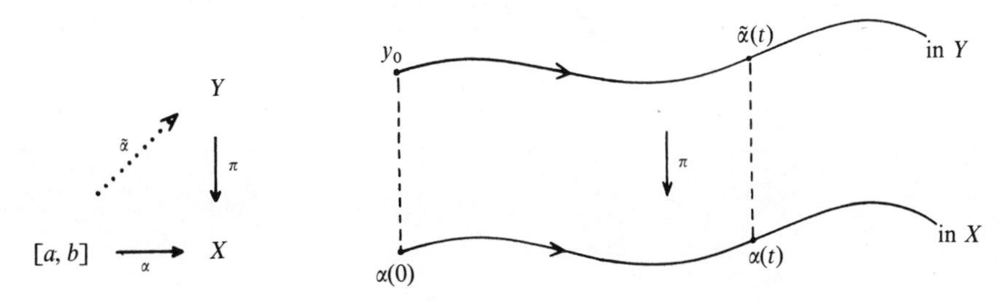

Lemma (Existence and Uniqueness of Path Liftings). *If Y is a covering space of X, then for every path α in X and every $y_0 \in Y$ over $\alpha(a)$ there is one and only one lifting $\tilde{\alpha}$ of α starting at y_0.*

Proof. If $U \subset X$ is open and $Y|U$ is trivial, it is easy to see exactly what are all the liftings of all paths β entirely contained in U: relative to $Y|U \cong U \times \Lambda$ they are exactly the paths $\tilde{\beta}_\lambda$ in Y defined by $\tilde{\beta}_\lambda(t) := (\beta(t), \lambda)$.

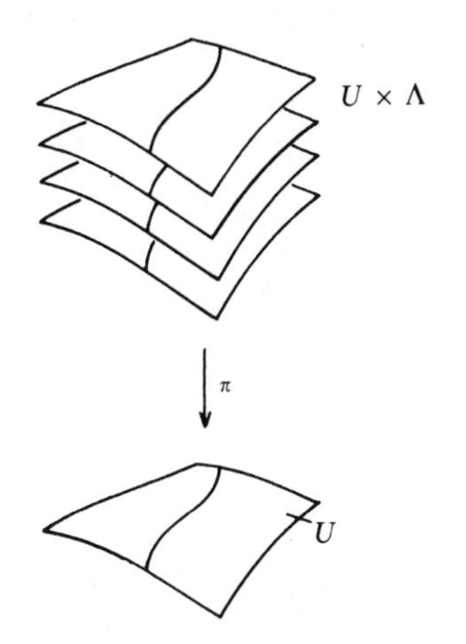

Now let α be given from $[0, 1]$ (no loss of generality) into X, and let y_0 lie over $\alpha(0)$.

Uniqueness. Let $\tilde{a}$ and $\hat{a}$ be two liftings starting at y_0. As shown by the reasoning for $Y|U$, the two subsets of $[0, 1]$ on which $\tilde{a}$ and $\hat{a}$ coincide (resp. do not coincide) are both open; the first is non-empty, since it contains 0, so it must be equal to $[0, 1]$ by connectedness.

Existence. The set of $\tau \in [0, 1]$ for which there is a lifting of $\alpha|[0, \tau]$ starting at y_0 is non-empty, since it contains 0. Let t_0 be the supremum of this set. Pick an open neighborhood U of $\alpha(t_0)$ over which Y is trivial, and an $\varepsilon > 0$ such that $[t_0 - \varepsilon, t_0 + \varepsilon] \cap [0, 1]$ is mapped by α entirely into U. Let $\tilde{\alpha}: [0, \tau] \to Y$ be a lifting of $\alpha|[0, \tau]$ for some $\tau \in [t_0 - \varepsilon/2, t_0] \cap [0, 1]$, and $\tilde{\beta}$ the lifting of $\alpha|[\tau, t_0 + \varepsilon/2] \cap [0, 1]$ starting at $\tilde{\alpha}(\tau)$.

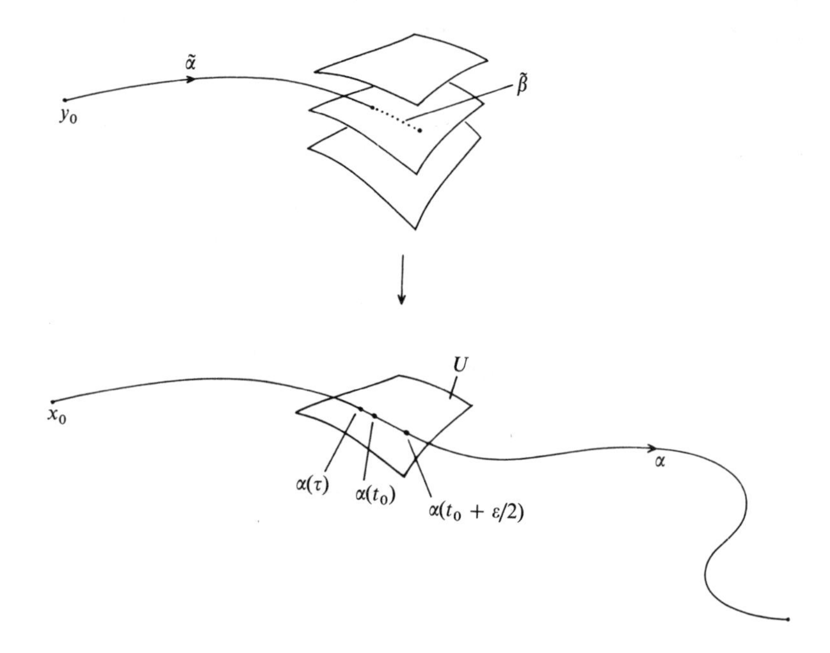

Then

$$\hat{\alpha}(t) := \begin{cases} \tilde{\alpha}(t) & \text{for } t \in [0, \tau], \\ \tilde{\beta}(t) & \text{for } t \in [\tau, t_0 + \varepsilon/2] \cap [0, 1] \end{cases}$$

defines a lifting of $\alpha|[0, b]$ starting at y_0, and moreover b is either equal to 1 (if $t_0 = 1$), in which case we are done, or greater than t_0. But the latter case cannot occur by definition of t_0, so $\hat{\alpha}$ is the desired lifting of α, qed. □

This lemma answers the two most obvious questions about path liftings. The other most important thing that must be known about path liftings is "continuous dependence on additional parameters". Let's imagine we're given not only a single path α, but a whole "family", i.e. a homotopy

$$h: Z \times [0, 1] \to X,$$

and correspondingly, instead of a single starting point y_0 over $\alpha(0)$ we have a whole "continuous starting point map" $\tilde{h}_0: Z \to Y$ over h_0, i.e. such that $\pi \circ \tilde{h}_0 = h_0$. Now if for each fixed $z \in Z$ we lift the corresponding path to a path starting at the prescribed point $\tilde{h}_0(z)$, we obtain in all a map

$$\tilde{h}: Z \times [0, 1] \to Y$$

over h. The question is whether $\tilde{h}$ can fail to be continuous:

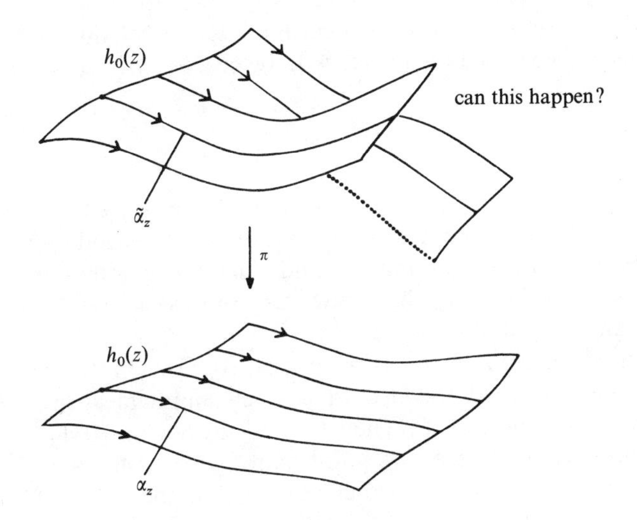

It can't! To prove it we'll again have to do a bit of punctilious work, but in return we'll have in our hands a very useful instrument for the theory of covering spaces.

Lemma (Lifting of Homotopies). *Let Y be a covering space of X, let Z be another topological space, $h: Z \times [0, 1] \to X$ a continuous map, and*

$$\tilde{h}_0: Z \to Y$$

a continuous "lifting" of h_0 ("prescribed lifting of the starting point map"):

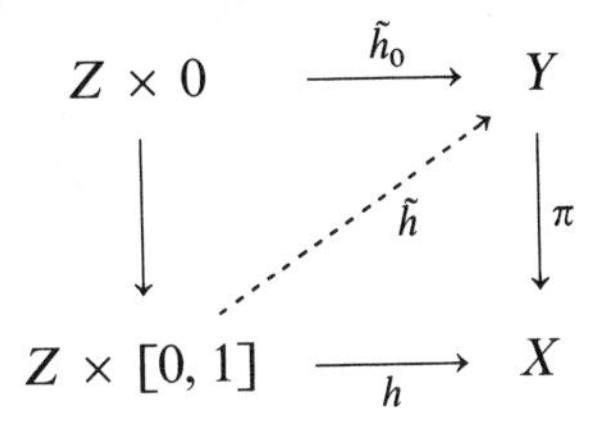

Then the map

$$\tilde{h}: Z \times [0, 1] \to Y,$$

$$(z, t) \to \tilde{\alpha}_z(t),$$

given by the lifting of the individual paths $\alpha_z: [0, 1] \to X, t \to h(z, t)$ *starting at* $\tilde{h}_0(z)$, *is continuous.*

PROOF. We will abbreviate the ε-neighborhood $(t_0 - \varepsilon, t_0 + \varepsilon) \cap [0, 1]$ of t_0 in $[0, 1]$ by $I_\varepsilon(t_0)$. An open box $\Omega \times I_\varepsilon(t_0)$ in $Z \times [0, 1]$ will be called "small" if it is mapped by h into an open set $U \subset X$ for which $Y|U$ is trivial. Now if $\tilde{h}$ is continuous on a "vertical line" $\Omega \times t_1$ of this box, then it is continuous in the whole box. In fact, with respect to a trivialization

$$Y|U \cong U \times \Lambda$$

the Λ-coordinate of $\tilde{h}|\Omega \times I_\varepsilon$ (which is what matters, since the U-coordinate is given by h, which is continuous anyway) does not depend on t, since each individual lifted path is continuous; and since $\tilde{h}$ is continuous on $\Omega \times t_1$, it must be continuous everywhere in $\Omega \times I_\varepsilon$. In this case we'll say the box is not only "small", but also "good".

Now for a fixed $z \in Z$ let T be the set of $t \in [0, 1]$ such that there is a small and good neighborhood box $\Omega \times I_\varepsilon(t)$, which simply means that $\tilde{h}_t$ is continuous in a neighborhood of z. Then T is trivially open in $[0, 1]$, and because of the continuity of the starting point map $\tilde{h}_0$ we have $0 \in T$. So now we only have to prove that T is also closed, because then $T = [0, 1]$ by connectedness, and consequently $\tilde{h}$ is continuous everywhere. So let $t_0 \in \bar{T}$. By continuity of h there is a "small" box $\Omega \times I_\varepsilon(t_0)$ around (z, t_0), and because $t_0 \in \bar{T}$ there is a $t_1 \in I_\varepsilon(t_0) \cap T$. Then $\tilde{h}_{t_1}$ is continuous on a neighborhood Ω_1 of z, hence $\tilde{h}$ is continuous on all of $(\Omega \cap \Omega_1) \times I_\varepsilon(t_0)$, and it follows that $t_0 \in T$, qed. □

As a first corollary we notice the

Monodromy Lemma. *Let Y be a covering space of X and α, β two paths in X which are homotopic with fixed endpoints, i.e. there is a homotopy*

$$h: [0, 1] \times [0, 1] \to X$$

between $h_0 = \alpha$ *and* $h_1 = \beta$ *with* $h_t(0) = \alpha(0)$ *and* $h_t(1) = \alpha(1)$ *for all* t. *Now if* $\tilde{\alpha}$ *and* $\tilde{\beta}$ *are liftings of* α *and* β *starting at the same point* y_0, *then they end at the same point as well*: $\tilde{\alpha}(1) = \tilde{\beta}(1)$.

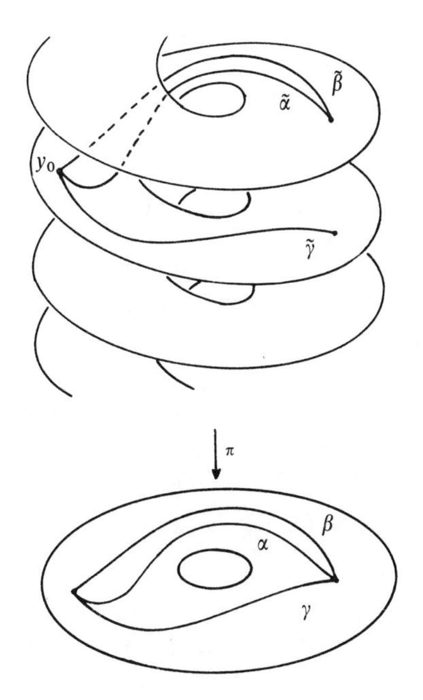

PROOF. Lifting each individual h_t to a path $\tilde{h}_t$ starting at y_0, the "ending point map" $t \to \tilde{h}_t(1)$ maps into the fiber over $\alpha(1)$; this we know independently of the lemma, because $\pi \circ \tilde{h}_t = h_t$. Now by the lemma this map is continuous, hence constant since the fiber is discrete, qed. □

§4. Introduction to the Classification of Covering Spaces

A covering space of X is at first a geometric object which cannot be immediately surveyed. If we want to get a general understanding, a classification of all covering spaces, we have to look around for "distinctive features" of coverings, i.e. we want to associate to every covering some data, some feature or mark of some sort that will be mathematically accessible and algebraical if possible, in such a way that two covering spaces of X are assigned the same features if and only if they are isomorphic. The classification of covering spaces up to isomorphism is thus reduced to the survey of the features, which hopefully simplifies the problem.

Many classification problems in mathematics are treated using this idea. A simple example that you all know is the classification of quadratic forms on an n-dimensional real vector space. A very obvious feature, an "invariant of isomorphism", of quadratic forms is their rank. But the rank does not characterize the isomorphism class, so one has to look for additional features, and one of them is for instance the signature, i.e. the difference $p - q$ of the maximal dimensions p and q of subspaces where the form is positive and negative definite, respectively. (Rank $p + q$ and signature $p - q$ evidently determine p and q, and conversely.) Then the classification theorem asserts ("Sylvester's law of inertia"): Two quadratic forms on V are isomorphic if and only if they have same rank and same signature. With this the surveying of isomorphism classes of quadratic forms is reduced to the surveying of all possible pairs (r, σ) of numbers which can occur as rank and signature of a form, which is of course a much simpler problem: The possible pairs are all those of the form $(p + q, p - q)$ with $0 \leq p, q$ and $p + q \leq n$.

For our covering spaces the *path-lifting behavior* yields such a distinctive feature. We restrict ourselves here to path-connected spaces, and we also consider in each space a given fixed point ("basepoint"). Of course covering maps are supposed to preserve basepoints, which we will write as

$$\pi : (Y, y_0) \to (X, x_0).$$

Path-connectedness, in view of the aims of the theory of coverings spaces, is not an essential limitation, and using a notation for the basepoint does not of course harm the mathematical content of the theory in the least.

Now we will say that two covering spaces (Y, y_0) and (Y', y'_0) of (X, x_0) have the same lifting behavior if for every two paths α and β in X from x_0 to some other point x_1 the following holds: The two lifted paths $\tilde{\alpha}$ and $\tilde{\beta}$ in Y end at the same point if and only if the two lifted paths $\tilde{\alpha}'$ and $\tilde{\beta}'$ in Y' starting at y'_0 also end at the same point.

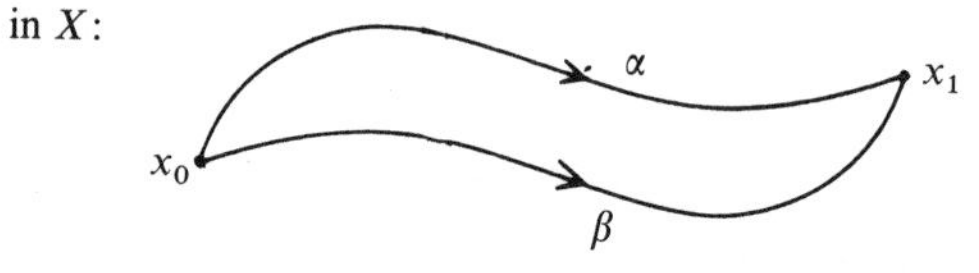

in X: α x_1 x_0 β

Now if this lifting behavior is to play for covering spaces a role similar to that played by rank and signature for quadratic forms, there are two very different questions that must be answerable:

(a) To what extent is a covering space determined by its lifting behavior? and

(b) How can the lifting behavior be "understood" algebraically?

The next two sections answer these questions, but I will take the perhaps unnecessary precaution of first delineating the principle on which the answer is based, so we won't get bogged down in the details.

For (a). Isomorphic coverings have of course the same lifting behavior; the question is whether the converse holds.

The lifting behavior decides about the liftability of continuous maps

$$f : (Z, z_0) \to (X, x_0)$$

from path-connected spaces Z in the following way: If a lifting $\tilde{f}$ exists in the first place,

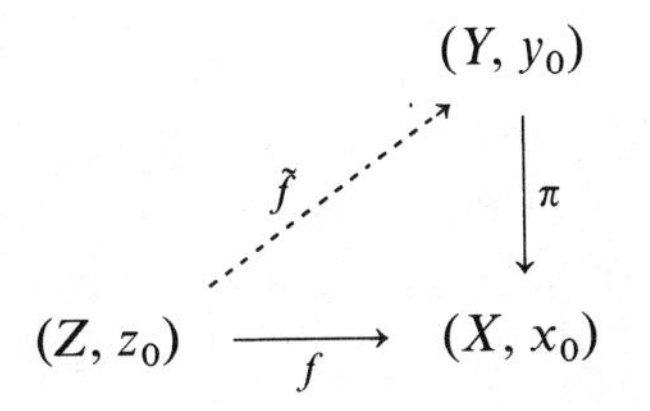

then of course for every path α from z_0 to z the path $\tilde{f} \circ \alpha$ is already the lifting starting at y_0 of the path $f \circ \alpha$, so $\tilde{f}(z)$ is the endpoint of the lifting of $f \circ \alpha$ starting at y_0:

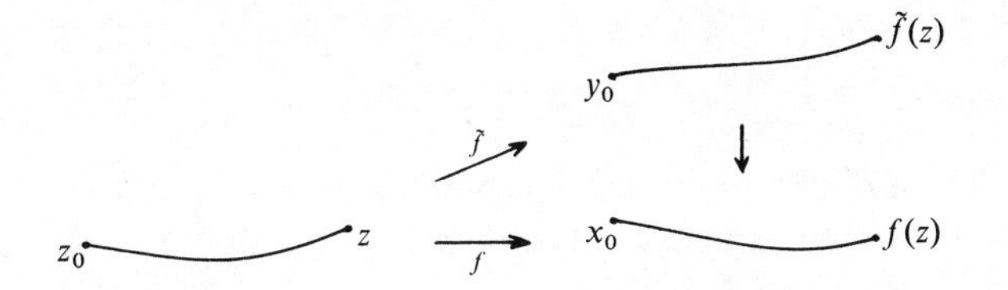

Next, it follows from the uniqueness of path liftings that for a given f there is at most one lifting $\tilde{f}$ with $\tilde{f}(z_0) = y_0$. But now we see also how such an $\tilde{f}$ is to be constructed, using path lifting, if we are given f to begin with: For $z \in Z$, choose a path α from z_0 to z, take its image, lift it and make the endpoint of the lifting equal to $\tilde{f}(z)$.

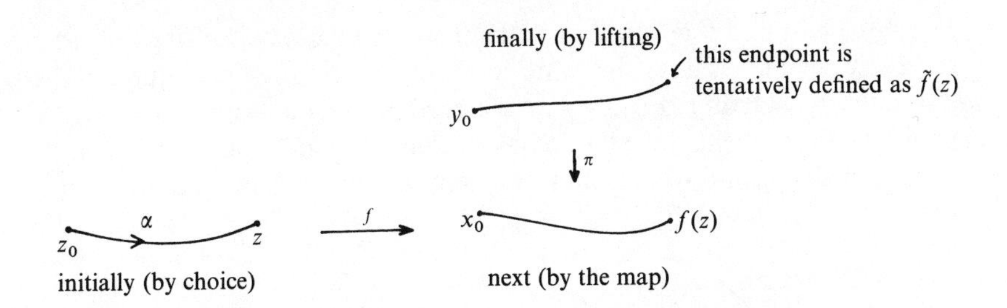

Now comes in a problem of "well-definedness": Is the $\tilde{f}(z)$ defined in this way independent of the choice of α?

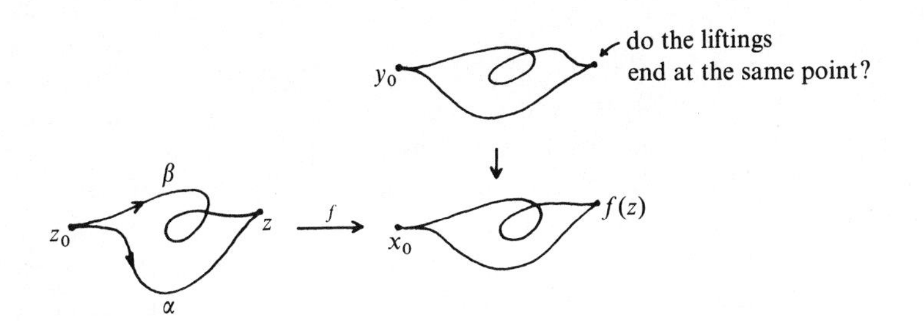

For this to be true we must have that for any two paths α and β that start at z_0 and end at the same point, the liftings of $f \circ \alpha$ and $f \circ \beta$ starting at y_0 must also end at the same point. Whether this condition, which is obviously necessary for the existence of $\tilde{f}$, is satisfied, is a consequence of the lifting behavior of the covering, and we will see (liftability criterion in §5) that the same condition, under appropriate assumptions, is actually sufficient for the existence of a continuous lifting $\tilde{f}$. In particular this "criterion" is clearly satisfied if the two given maps are covering maps with the same lifting behavior:

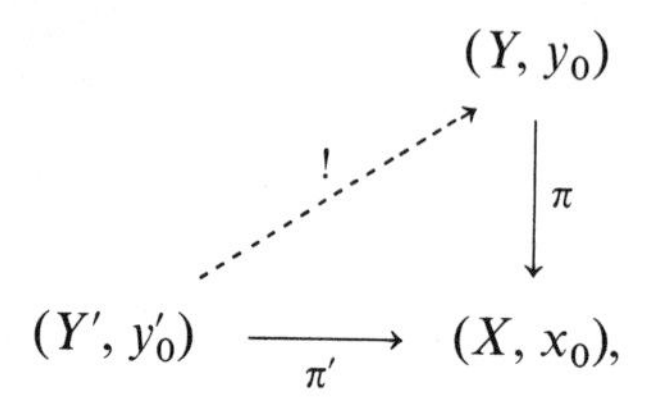

and another application of the argument, switching the two covering spaces, gives us isomorphisms between the spaces that are inverse to each other:

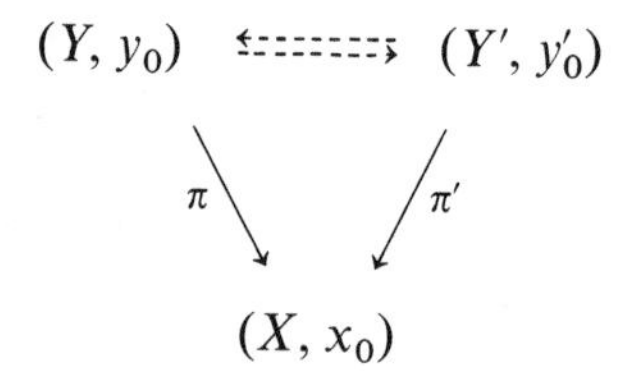

For (b). Let α and β be the two paths from x_0 to x. Their liftings $\tilde{\alpha}$ and $\tilde{\beta}$ starting at y_0 evidently have the same endpoint if and only if the lifting of the closed path $\alpha\beta^-$ starting at y_0 is again a closed path, or a "loop". So in order to know the lifting behavior all one has to know is which loops at x_0 can be lifted to loops at y_0 and which can't. By the monodromy lemma this depends

$$\alpha\beta^- : t \mapsto \begin{cases} \alpha(2t) & 0 \leq t \leq \frac{1}{2} \\ \beta(2-2t) & \frac{1}{2} \leq t \leq 1 \end{cases}$$

only on the homotopy class (with fixed endpoints, both equal to x_0). But the set of homotopy classes of loops at x_0 forms a groups in a canonical way, namely, the so-called fundamental group $\pi_1(X, x_0)$, and the subset of classes of loops that can be lifted to other loops forms a subgroup $G(Y, y_0)$ of $\pi_1(X, x_0)$. Thus knowing the lifting behavior means knowing this subgroup: the subgroup is the algebraic "feature" we associate to a covering space. So the classification of covering spaces consists of the uniqueness theorem mentioned in part (a), that says that two coverings are isomorphic if and only if they have the same lifting behavior, hence the same group G; and an existence theorem, yet to be formulated, which indicates to what extent there really is, for a given subgroup $G \subset \pi_1(X, x_0)$, a covering space with $G(Y, y_0) = G$. The program outlined here shall now be carried out in detail in §§5 and 6.

§5. Fundamental Group and Lifting Behavior

Definition (Category of Spaces with Basepoint). By a space with basepoint we simply mean a pair (X, x_0) consisting of a topological space X and a point $x_0 \in X$. A continuous, basepoint-preserving map $f: (X, x_0) \to (Y, y_0)$ is, as the name says, a continuous map $f: X \to Y$ with $f(x_0) = y_0$. In particular we mean by a covering map $\pi: (Y, y_0) \to (X, x_0)$ a covering map $\pi: Y \to X$ with $\pi(y_0) = x_0$.

Definition (Fundamental Group). Let (X, x_0) be a space with basepoint. Let $\Omega(X, x_0)$ be the set of paths in X that start and end at x_0 ("loops at x_0"), and let $\Omega \times \Omega \to \Omega, (\alpha, \beta) \to \alpha\beta$ be the composition map given by $\alpha\beta(t) := \alpha(2t)$ for $0 \leq t \leq \frac{1}{2}$ and $\beta(2t - 1)$ for $\frac{1}{2} \leq t \leq 1$ respectively. Then on the set $\pi_1(X, x_0) := \Omega(X, x_0)/\simeq$ of homotopy classes ($\simeq$ denotes here homotopy with endpoints fixed and equal to x_0) there is a well-defined composition law given by $[\alpha][\beta] := [\alpha\beta]$ which makes $\pi_1(X, x_0)$ into a group. This group is called the *fundamental group* of the space with basepoint (X, x_0).

I'll skip the proofs of the various steps necessary to justify these assertions (well-definedness of the composition map, associativity, existence of 1 and of the inverse), which are very simple. The reader who hasn't had any contact with path homotopy and wants to see how these things work might profit by going through pages 78–88 of [12].

Notation. Evidently, this construction actually gives us in a canonical way a covariant functor π_1 from the category of spaces with basepoint and basepoint-preserving continuous maps into the category of groups and homomorphisms, namely $\pi_1 f: \pi_1(X, x_0) \to \pi_1(Y, y_0), [\alpha] \to [f \circ \alpha]$. But instead of $\pi_1 f$ we will use the common notation f_*.

Corollary of the Monodromy Lemma (Behavior of the Fundamental Group Relative to Covering Maps). *If $\pi:(Y, y_0) \to (X, x_0)$ is a covering map, then the induced group homomorphism $\pi_*: \pi_1(Y, y_0) \to \pi_1(X, x_0)$ is injective.*

Proof. Let $\pi_*[\gamma] = 1 \in \pi_1(X, x_0)$. Then there is a homotopy h with fixed endpoints x_0 between $\pi \circ \gamma$ and the constant path $[0, 1] \to \{x_0\}$. Now we lift h to a homotopy $\tilde{h}$ with fixed endpoints y_0, and then $\tilde{h}_0 = \gamma$ and $\tilde{h}_1$ is a lifting of the constant path, hence constant. So $[\gamma] = 1 \in \pi_1(Y, y_0)$, qed. □

Definition (Characteristic Group of a Covering Space). Let $\pi: (Y, y_0) \to (X, x_0)$ be a covering map. Then call the image of the injective homomorphism $\pi_*: \pi_1(Y, y_0) \to \pi_1(X, x_0)$ the characteristic subgroup of the covering, and denote it by $G(Y, y_0) \subset \pi_1(X, x_0)$.

For a loop at x_0 to be liftable to a loop at y_0 just means of course that it is the projection of a loop at y_0; and so the group $G(Y, y_0) \subset \pi_1(X, x_0)$ is the subgroup announced in the last section, which contains all the information about the lifting behavior of the covering.

Now if $f: (Z, z_0) \to (X, x_0)$ is continuous and $\tilde{f}: (Z, z_0) \to (Y, y_0)$ is a lifting of f, i.e. $\pi \circ \tilde{f} = f$

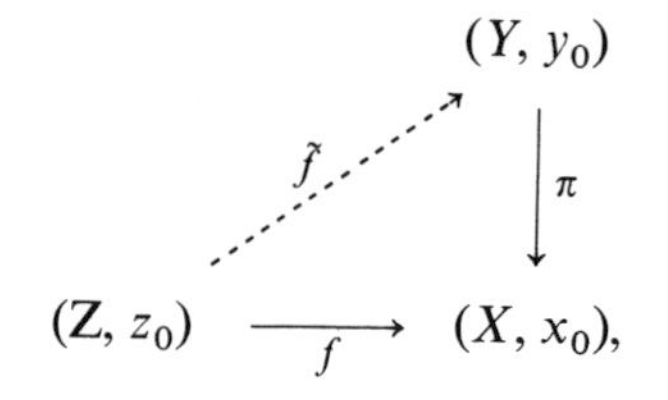

we obviously have $f_* \pi_1(Z, z_0) = \pi_*(\tilde{f}_* \pi_1(Z, z_0)) \subset \pi_* \pi_1(Y, y_0) = G(Y, y_0)$, which means that $f_* \pi_1(Z, z_0) \subset G(Y, y_0)$ is a necessary condition for the liftability of a given map $f: (Z, z_0) \to (X, x_0)$. In order to be able to formulate to what extent this condition is also sufficient, I must first introduce yet another notion of connectedness, namely

Definition (Locally Path-Connected). A topological space is called locally path-connected if every neighborhood contains a path-connected neighborhood.

Remark. Manifolds are locally path-connected (clear), and so are CW-complexes; see the (more general) theorem in [16], III, 3.6.

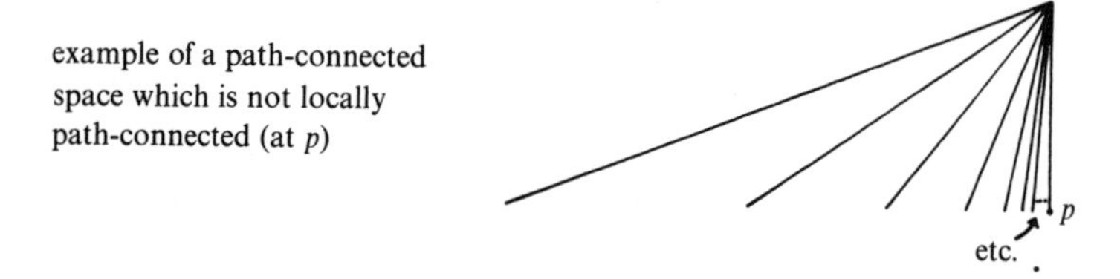

Note. *In a locally path-connected space every neighborhood V of a point p actually contains an* open *path-connected subneighborhood, for instance the set of all $x \in V$ which can be reached by a path in V starting at p.*

Now then to the

Liftability Criterion. Let $\pi: (Y, y_0) \to (X, x_0)$ be a covering map, Z a path-connected and locally path-connected space, and $f: (Z, z_0) \to (X, x_0)$ a continuous map. There exists a lifting $\tilde{f}: (Z, z_0) \to (Y, y_0)$ of $\tilde{f}$ (which is unique) if and only if f_* maps the fundamental group $\pi_1(Z, z_0)$ into the characteristic subgroup $G(Y, y_0) \subset \pi_1(X, x_0)$ of the covering:

$$\begin{array}{c} \pi_1(Y, y_0) \\ \cong \downarrow \pi_* \\ \pi_1(Z, z_0) \to G(Y, y_0) \subset \pi_1(X, x_0). \end{array}$$

By the way, here we have a simple example of what I had called in Chapter V, §7 the "second main reason for the usefulness of the notion of homotopy" (see p. 74): The geometric problem is solvable if and only if the corresponding algebraic problem, which arises by applying the fundamental group functor π_1 to this situation, is solvable. A lifting $\tilde{f}$ exists if and only if f_* is liftable to a group homomorphism φ:

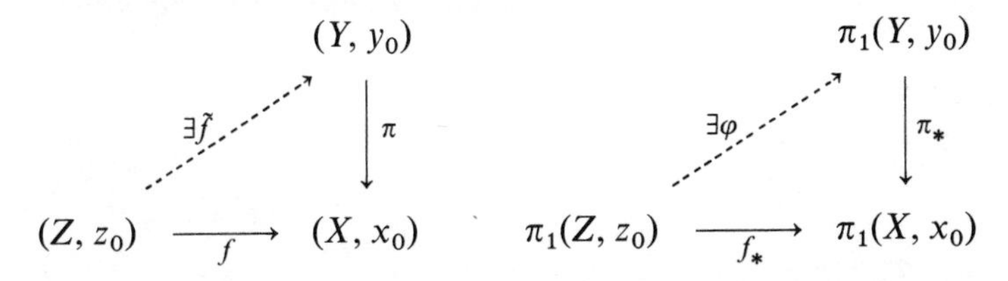

Proof. The condition is obviously necessary, and because of the path-connectedness of Z and the uniqueness of liftings of paths there can be at most one $\tilde{f}$. Now suppose f satisfies the condition. For $z \in Z$ we define $\tilde{f}(z)$ in the way already described in §4, namely: We take a path α from z_0 to z, then lift $f \circ \alpha$ to a path starting at y_0, and define $\tilde{f}(z)$ as the endpoint of this lifted path in Y. This endpoint does not depend on the choice of α, because if β is another path from z_0 to z, then the loop $(f \circ \alpha)(f \circ \beta)^-$ represents an element of $G(Y, y_0)$, so it can be lifted to a loop in Y, and consequently the liftings of $f \circ \alpha$ and $f \circ \beta$ end at the same point, which is exactly the well-defined $\tilde{f}(z)$. Then of course $\pi \circ \tilde{f} = f$, and all that remains to be shown is that $\tilde{f}$ is continuous. Here comes in the local path-connectedness of Z:

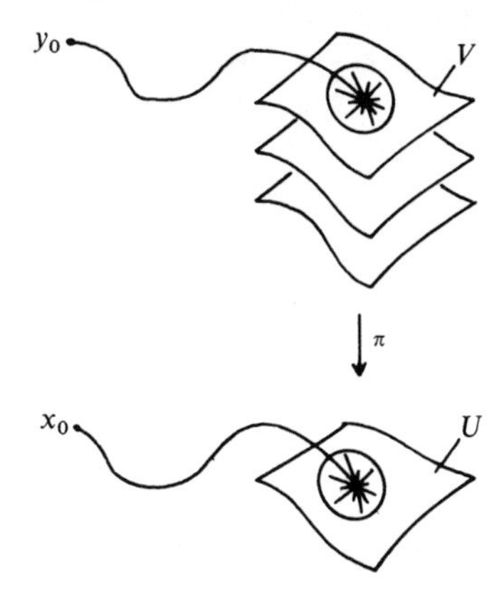

Let $z_1 \in Z$ and let V be an open neighborhood of $\tilde{f}(z_1)$ in Y. Without loss of generality, we can assume V is so small that $\pi | V$ is homeomorphism on the open set $U := \pi(V)$. Now choose a path-connected neighborhood W of z_1 so small that $f(W) \subset U$. To connect the points $w \in W$ with z_0, we choose a fixed path α from z_0 to z_1 and add to it small "stitch paths", entirely contained in W, from z_1 to the w. Then it is evident that $\tilde{f} | W = (\pi | V)^{-1} \circ f | W$, and in particular $\tilde{f}(W) \subset V$, qed. □

§6. The Classification of Covering Spaces

Uniqueness Theorem. *Between two path-connected and locally path-connected covering spaces (Y, y_0) and (Y', y_0') of (X, x_0) there is a basepoint-preserving isomorphism if and only if they have the same characteristic subgroup*

$$G(Y, y_0) = G(Y', y_0') \subset \pi_1(X, x_0).$$

Proof. Necessity is clear: If $\varphi: (Y, y_0) \to (Y', y_0')$ is such an isomorphism, then

$$\begin{aligned} G(Y, y_0) = \pi_* \pi_1(Y, y_0) = (\pi' \circ \varphi)_* \pi_1(Y, y_0) &= \pi'_*(\varphi_* \pi_1(Y, y_0)) \\ &= \pi'_* \pi_1(Y', y_0') = G(Y', y_0'). \end{aligned}$$

Conversely, if the condition is satisfied, then we can lift the two projections to one another, by the liftability criterion:

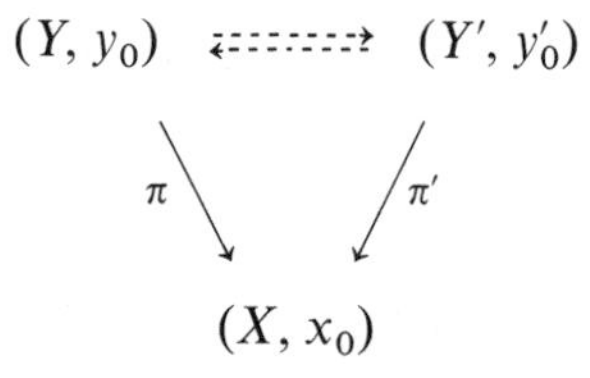

$$\begin{array}{ccc} (Y, y_0) & \rightleftarrows & (Y', y_0') \\ & {\scriptstyle\pi}\searrow \quad \swarrow{\scriptstyle\pi'} & \\ & (X, x_0) & \end{array}$$

and then the compositions of these two liftings are liftings of π and π' to themselves, hence equal to the identity on Y and Y', respectively, by the uniqueness of liftings. qed. □

Existence Theorem? In the uniqueness theorem we have assumed the covering spaces, and consequently the base space X, to be path-connected and locally path-connected. So from now on we formulate the existence question accordingly: Let (X, x_0) be a path-connected and locally path-connected space, and $G \subset \pi_1(X, x_0)$ an arbitrary subgroup of the fundamental group. Question: Is there a path-connected covering space (Y, y_0) of (X, x_0) (local path-connectedness is transferred from X to Y anyway), and with $G(Y, y_0) = G$? Now this is *not* true in such generality. Why not? Do we have to make additional assumptions about X? Or about G? And what would they be?

Instead of presenting the theorem immediately, I'd like to apply again here the inductive method of exposition, which is very instructive, though unfortunately too time-consuming to be used all the time; namely, simulating the situation, so characteristic of a mathematician's life, where a theorem is not only to be proved, but has first of all to be found.

The first stage of a theorem is generally made up of *wishes*, as they present themselves naturally when we get sufficiently acquainted with some matter. Theorems then follows when we try to prove the desired assertions, analyze the difficulties we find, and seek to remedy them by using extra assumptions, which we try to make as weak as possible. In our case, for instance, we wish that there always be a covering space whose characteristic subgroup is the prescribed G. So let's try to prove it!

PROOF OF THE AS YET UNFOUND EXISTENCE THEOREM. First of all we must create Y as a set. If we already had a covering map $(Y, y_0) \to (X, x_0)$ as desired—in what way could we characterize the points of Y_x by means of objects expressible in terms of (X, x_0) and G? Well, for every path α from x_0 to x there would correspond a perfectly determined point of the fiber Y_x over x, namely the point where the lifting of α that starts at y_0 ends. All points of Y_x would be obtained in this way, and two paths α, β would determine the same point in Y_x if and only if the loop $\alpha\beta^-$ represented an element of G. So if Y did not yet exist, how would one proceed to create it? Like this:

Definition. Let $\Omega(X, x_0, x)$ be the set of paths from x_0 to x. We define on this set an equivalence relation by $\alpha \sim \beta :\Leftrightarrow [\alpha\beta^-] \in G$, and define the sets Y_x and Y by

$$Y_x := \Omega(X, x_0, x)/\sim,$$

$$Y := \bigcup_{x \in X} Y_x.$$

Moreover, let y_0 be the equivalence class of the constant path in $\Omega(X, x_0, x_0)$ and let $\pi\colon Y \to X$ be given by $Y_x \to \{x\}$. Then in any case π is a surjective map between sets and we have $\pi(y_0) = x_0$.

So now our task consists "only" in providing Y with a topology that will make it into a path-connected space and make $(Y, y_0) \to (X, x_0)$ into a covering map with $G(Y, y_0) = G$.

When we think about our construction in geometric terms, we see immediately that besides the set Y and the map $\pi\colon (Y, y_0) \to (X, x_0)$ we already have something else in our hands: path lifting. For $\alpha \in \Omega(X, x_0, x)$ we will denote the $\sim$ equivalence classes by $[\alpha]_\sim$, to avoid confusion with the homotopy class $[\alpha]$. For $t \in [0, 1]$ let $\alpha_t \in \Omega(X, x_0, \alpha(t))$ denote the "partial path" defined by $s \mapsto \alpha(ts)$.

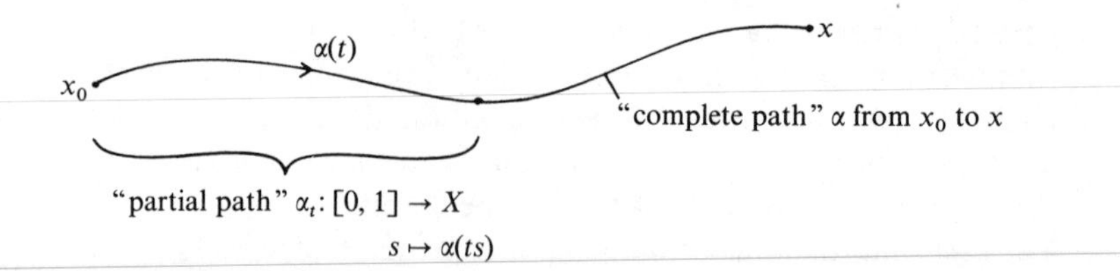

The intuition underlying our construction points naturally to a topology relative to which the lifting of α is already given by $\tilde{\alpha}(t) := [\alpha_t]_\sim$. If we follow it, we find automatically how the topology on Y must be defined. To this purpose we introduce the following notation: For a path α from x_0 to x and an open path-connected neighborhood U of x, denote by $V(U, \alpha) \subset Y$ the set of equivalence classes $[\alpha\beta]_\sim$ of paths which can be obtained by combining α with paths β in U that start at x:

Because of the local path-connectedness of X, these U form a neighborhood basis of x, and consequently the $V(U, \alpha)$ should obviously form a neighborhood basis of $y \in Y$ in the topology we're choosing. But before we make this into a definition, let's observe that $V(U, \alpha)$ depends only on $y = [\alpha]$ and U, but not on the choice of the representative path α:

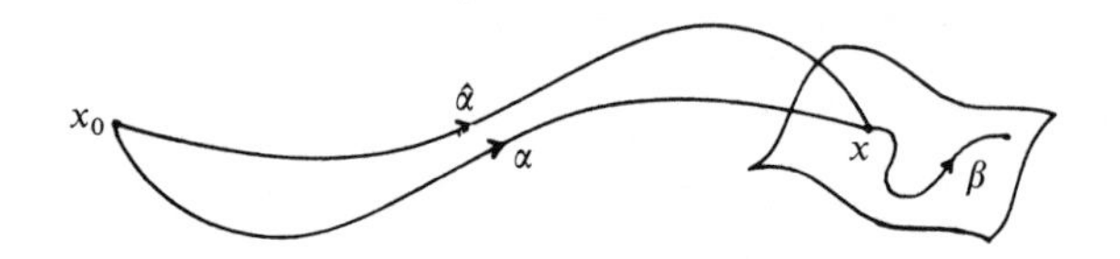

From $\hat{\alpha} \sim \alpha$ it follows $[(\alpha\beta)(\hat{\alpha}\beta^-)] = [\alpha\beta\beta^-\hat{\alpha}^-] = [\alpha\hat{\alpha}^-] \in G$, and so $[\alpha\beta]_\sim = [\hat{\alpha}\beta]_\sim$. Because of this independence from α we can write $V(U, y)$ instead of $V(U, \alpha)$, which we will proceed to do. Notice that $\pi(V(U, y)) = U$.

Definition. Call $V \subset Y$ open if for every $y \in V$ there is an open path-connected neighborhood U of $\pi(y)$ such that $V(U, y) \subset V$.

So now our task is to check, and if necessary to guarantee using additional assumptions as weak as possible, that the following hold:

(a) $\varnothing$, Y are open;
(b) arbitrary unions of open sets are open;
(c) finite intersections of open sets are open;
(d) π is continuous;
(e) fibers are discrete;

(f) $\pi: Y \to X$ is locally trivial;
(g) Y is path-connected;
(h) $G(Y, y_0) = G$.

(a)–(d) are trivially satisfied: It's really a topology, and π is continuous. I'd like to remark here en passant that so far we haven't "given up" anything: If there is a covering space (Y', y_0') with the desired properties in the first place, then our construction too must have properties (e)–(h), since one can easily establish a homeomorphism between (Y, y_0) and (Y', y_0') over (X, x_0).

(e) *Discreteness of the Fibers.* Discreteness of the fibers is equivalent to the fact that for each $y \in Y_x$ there is a path-connected open neighborhood U of x such that y is the only point of $Y_x \cap V(U, y)$. What does this uniqueness mean? If $y = [\alpha]_\sim$, the other points of $Y_x \cap V(U, y)$ are exactly the $[\alpha\beta]_\sim$, where β is a loop in U based at x, and so we must find an U such that $[\alpha]_\sim = [\alpha\beta]_\sim$ for all loops β of this form.

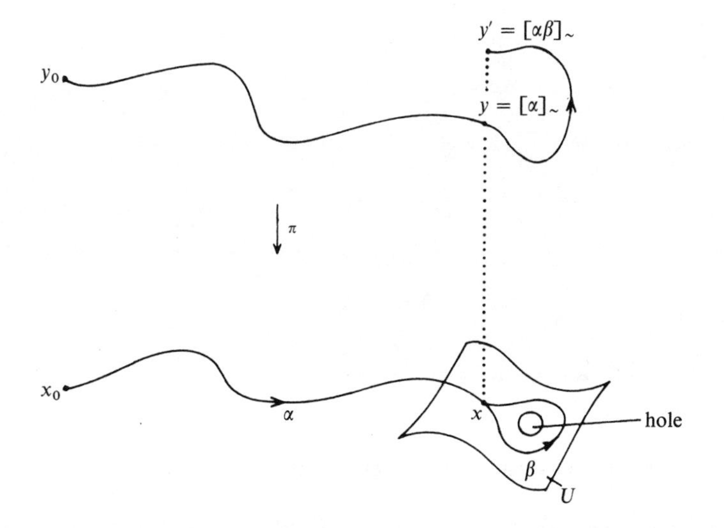

And here our ship runs aground, because without further assumptions about X the homotopy class $[\alpha(\alpha\beta)^-]$ has no reason to belong to G. If we think for example of the case $x = x_0$, $\alpha = \text{const}$, $G = \{1\}$, for which everything should work: for this case our condition means straight away that the loop β is null-homotopic in X. But there doesn't have to be any neighborhood U for which all loops in U based at x are null-homotopic in the big space X:

Example:

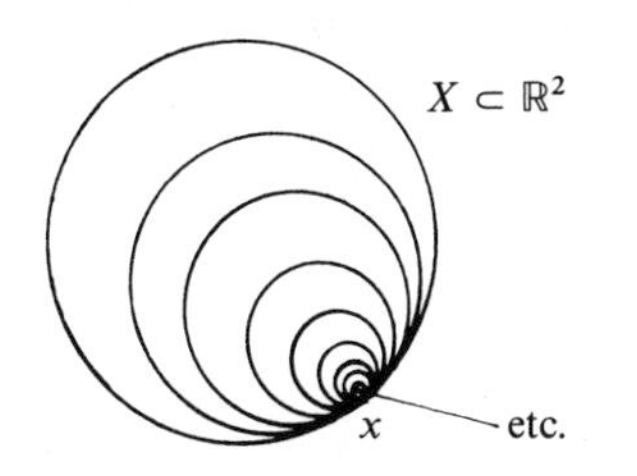

To get moving again, we simply assume that X has the desired property—there is nothing else we can do, otherwise the existence theorem is false already for the case $G = 1$. So:

Definition. A topological X is called semi-locally simply connected if every $x \in X$ possesses a neighborhood U such that every path contained in U and based at x is null-homotopic in X.

The condition is called *semi*-local because although the loops are "local", i.e. contained in U, their homotopies to the constant path are global, i.e. they are allowed to run in X. It is clear that this property is transferred from U to its subneighborhoods. "Locally simply connected" would be defined as follows: Every neighborhood contains a simply connected subneighborhood, i.e. a neighborhood V such that all loops inside V are null-homotopic there. The cone over the example above

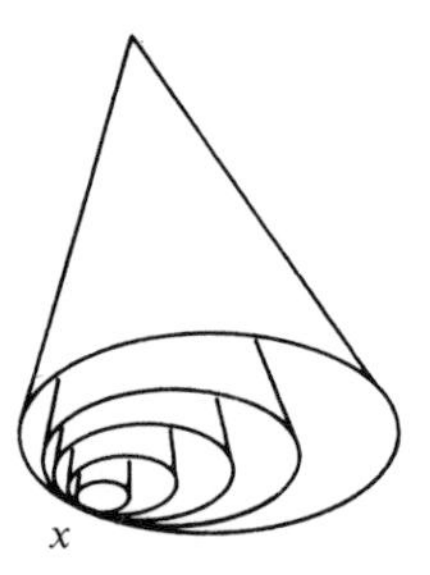

is semi-locally simply connected, but not locally simply connected. But this is beside the point; more important is the

Observation. Manifolds (clear) and also CW-complexes are always semi-locally (even locally) simply connected—see [16], III.3.6.

Additional Assumption. In the remainder of this "proof of the yet unfound existence theorem" X will be assumed to be semi-locally simply connected.

For small enough U we then have $Y_x \cap V(U, y) = \{y\}$, whence the discreteness of the fibers follows, (e)-qed.

(f) *Local Triviality.* Let $x \in X$ and U be an open path-connected neighborhood in which every loop at x is null-homotopic within the whole space X. Then it can be immediately verified that $V(U, y) = V(U, z)$ for every $z \in V(U, y)$; for $y \in Y_x$ the $V(U, y)$ are pairwise disjoint open sets and $\pi^{-1}(U) = \bigcup_{y \in Y_x} V(U, y)$. Then the projection π and the correspondence $V(U, y) \to \{y\}$, well-defined for $y \in Y_x$, together define a continuous bijective map $h: \pi^{-1}(U) \to U \times Y_x$ over U, which we still have to check for openness (i.e. that it takes open sets into open sets). To do this it suffices to verify that the projection itself is open, which in its turn is very easy. The sets $V(U, y)$, for $y \in Y$ and open path-connected sets $U \subset X$, form a basis of the topology in Y, so we only have to know that $\pi(V(U, y))$ is open; but this set is equal to U. So h is open and $\pi: Y \to X$ is indeed locally trivial. (f)-qed.

(g) *Path-connectedness of* Y. If $y = [\alpha]_\sim$, then $t \mapsto [\alpha_t]_\sim$ does indeed give a path from y_0 to y, where $\alpha_t(s) := \alpha(st)$. (g)-qed.

(h) $G(Y, y_0) = G$. A loop α at x_0 represents an element of $G(Y, y_0)$ if and only if it can be lifted to a loop at y_0. But this is the case if and only if (see (g)) $[\alpha]_\sim = y_0$, i.e. $[\alpha]_\sim = [x_0]_\sim$, and thus if and only if $[\alpha x_0] = [\alpha] \in G$. (h)-qed.

So with the one additional assumption we made along the way everything goes through, and we have proved:

Existence Theorem. *If X is path-connected, locally path-connected and semi-locally simply connected, and if $G \subset \pi_1(X, x_0)$ is an arbitrary subgroup, then there is a path-connected and locally path-connected covering space (Y, y_0) of (X, x_0) with $G(Y, y_0) = G$.*

Note. *By the monodromy lemma this clearly implies Y is semi-locally simply connected as well.*

§7. Covering Transformations and Universal Cover

The liftability criterion and the existence and uniqueness theorems form the core of covering space theory. We will now note some useful consequences.

Definition (Covering Transformation). By a covering transformation, or deck transformation, of a covering map $\pi: Y \to X$ we simply mean an automorphism of the covering, i.e. a homeomorphism $\varphi: Y \xrightarrow{\cong} Y$ over X:

$$\begin{array}{ccc} Y & \xrightarrow[\cong]{\varphi} & Y \\ & {}_{\pi}\searrow \quad \swarrow_{\pi} & \\ & X & \end{array} \quad \text{commutes.}$$

Covering transformations evidently form a group, which we will denote by $\mathscr{D}$.

As an immediate corollary of the uniqueness theorem we have the

Remark. Let Y be a path-connected and locally path-connected covering space of X and let $y_0, y_1 \in Y$ be points over $x_0 \in X$. There is a covering transformation $\varphi: Y \to Y$ with $\varphi(y_0) = y_1$ (which is unique) if and only if (Y, y_0) and (Y, y_1) have the same characteristic subgroup in $\pi_1(X, x_0)$.

In particular the group of covering transformations operates freely on Y: Only the identity has fixed points.

But what does the condition $G(Y, y_0) = G(Y, y_1)$ mean? To study this, we will take a look at the connection between $G(Y, y_0)$ and $G(Y, y_1)$ in general, for two points y_0 and y_1 over x_0. Let γ be any path in Y from y_0 to y_1, and let $\alpha := \pi \circ \gamma$ be its projection:

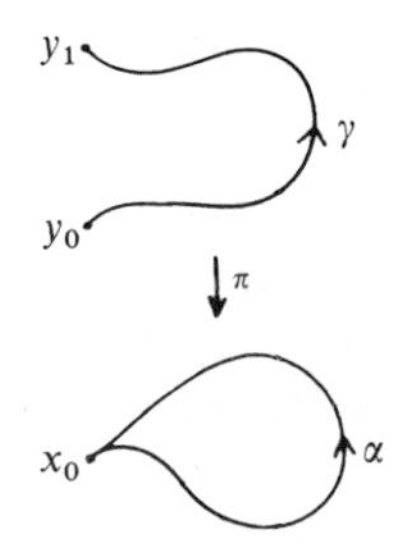

Then we have a commutative diagram of group isomorphisms,

$$\begin{array}{ccc}
\pi_1(Y, y_0) & \xrightarrow[\cong]{[\gamma^- \cdots \gamma]} & \pi_1(Y, y_1) \\
\cong \Big\downarrow \pi_* & & \cong \Big\downarrow \pi_* \\
G(Y, y_0) & \xrightarrow[{[\alpha]^{-1} \cdots [\alpha]}]{\cong} & G(Y, y_1).
\end{array}$$

and so $G(Y, y_1) = [\alpha]^{-1} G(Y, y_0)[\alpha]$, which means $G(Y, y_0) = G(Y, y_1)$ if and only if $[\alpha]$ is in the *normalizer* of $G(Y, y_0)$ in $\pi_1(X, x_0)$:

Review (Algebra). If B is a subgroup of a group A, the set

$$N_B := \{\alpha \in A \mid aBa^{-1} = B\}$$

is called the normalizer of B in A. The normalizer is itself a subgroup of A, and B is obviously normal in its normalizer: $B \lhd N_B \subset A$; the normalizer is simply the biggest group between B and A in which B is normal.

Theorem About the Group of Covering Transformations. *Let* $(Y, y_0) \to (X, x_0)$ *be a covering of a path-connected and locally path-connected space, and let* $G := G(Y, y_0)$ *be its characteristic subgroup, i.e. the image of the injective* *homomorphism $\pi_1(Y, y_0) \to \pi_1(X, x_0)$ induced by the projection. Then for every element $[\alpha] \in N_G$ of the normalizer of G in $\pi_1(X, x_0)$ there is exactly one covering transformation $\varphi_{[\alpha]}$ which maps y_0 into the endpoint $\tilde{\alpha}(1)$ of the lifting of α starting at y_0. Moreover, in this way we get a map $N_G \to \mathscr{D}$ which in fact defines a group isomorphism $N_G/G \xrightarrow{\cong} \mathscr{D}$.*

The proof I recommend as a pleasant exercise for you to get better acquainted with the many new concepts introduced in this chapter. Don't forget to include $\varphi_{[\alpha\beta]} = \varphi_{[\alpha]} \circ \varphi_{[\beta]}$ in the list of individual assertions to be proved. (The formula is correct as it stands, and not the other way around: although in $\alpha\beta$ the path α is described first, in $\varphi_{[\alpha]} \circ \varphi_{[\beta]}$ the covering transformation $\varphi_{[\beta]}$ is applied first.)

Corollary and Definition (Normal Coverings). The group of covering transformations of a path-connected and locally path-connected covering $Y \to X$ operates transitively on the fibers (i.e. for every two points of a fiber there is a covering transformation that takes one into the other; or, alternatively, the fibers are the orbits of the $\mathscr{D}$-action on Y) if and only if for some (and hence every) point $y_0 \in Y$ the group $G(Y, y_0)$ is normal in $\pi_1(X, \pi(y_0))$. Such coverings are called *normal* coverings.

Corollary. *For normal coverings $(Y, y_0) \to (X, x_0)$ the following hold:*

(i) $\mathscr{D} = \pi_1(X, x_0)/G(Y, y_0)$.

(ii) *The number of leaves of the covering is equal to the order of $\mathscr{D}$ (because the fibers are the orbits of the free $\mathscr{D}$-action), and hence also equal to the order of π_1/G, called in group theory the "index" of $G(Y, y_0)$ in $\pi_1(X, x_0)$.*

(iii) *The bijective map from the orbit space $Y/\mathscr{D}$ onto X defined by the projection $\pi: Y \to X$ is in fact a homeomorphism:*

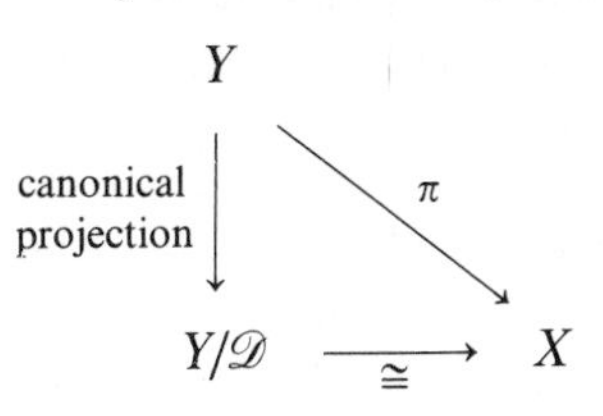

Proof of (iii). That the map is continuous follows from III, §2, note 1; that it is also open follows from the fact that $\pi: Y \to X$ is a local homeomorphism, hence open.

*

In particular all this is true for the case $G(Y, y_0) = \{1\}$, which we will study now. Since $\pi_1(Y, y_0) \overset{\pi_*}{\cong} G(Y, y_0)$ this case occurs if and only if the fundamental group of Y is trivial, and let's recall that such spaces are called simply connected:

Definition (Simply Connected). A path-connected space Y is called simply connected if for some (and hence all) $y_0 \in Y$ the fundamental group $\pi_1(Y, y_0)$ is trivial.

Thus the condition means that every loop in Y is null-homotopic. Contractible spaces are evidently simply connected, but so is for instance the sphere S^n for $n \geq 2$. Surprisingly enough, this fact isn't really entirely obvious. How so? Isn't it enough, given a loop α at q in S^n, to choose a point $p \in S^n$ outside the image of α, and use the fact that $\{q\}$ is a strong deformation retract of $S^n \backslash p$?

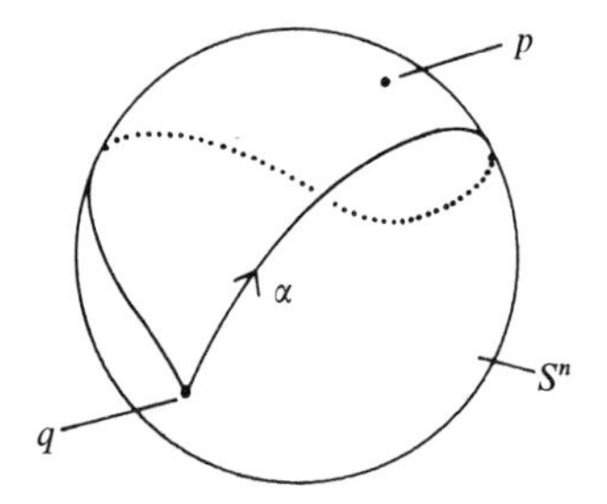

Yes, yes; but you must probably have already heard about "space-filling curves" (Peano, G., "Sur une courbe qui remplit toute une aire plane", *Math. Annalen* **36** (1890), 157–160), and by the same token there are sphere-filling loops, for which such a point p is not at all available.

A direct proof is not really deep, however: All one has to do is subdivide the interval $[0, 1]$ at points $0 = t_0 < \cdots < t_n = 1$, so finely that α is not "sphere-filling" on any of the subintervals; this is always possible because of continuity. Then the contractibility of $S^n \backslash pt$ implies there is a homotopy of α which fixes a at the endpoint t_i of the subintervals, and in addition takes α into a loop which maps each subinterval into some great circle. Now this curve is not sphere-filling, and the argument previously envisaged works.

If Y is a path-connected and locally path-connected covering space of a simply connected space X, then $Y \to X$ must have multiplicity one and so must be a homeomorphism. This observation is often very useful; it says that a simply connected space cannot have any interesting coverings—an immediate consequence of the uniqueness theorem. But now we want to consider coverings in which the covering space Y, and not the base space X, is simply connected.

Definition (Universal Cover). A path-connected and locally path-connected covering space $Y \to X$ is called a *universal cover* if Y is simply connected.

Corollary of the Classification of Covering Spaces. *If X is path-connected, locally path-connected and semi-locally simply connected, and if $x_0 \in X$, there* *is exactly one universal cover* $(\tilde{X}, \tilde{x}_0) \to (X, x_0)$, *up to a uniquely determined isomorphism.*

In this sense one could then speak of "the" universal cover $\tilde{X}$ of X.

What is so "universal" about the universal cover? A preliminary consideration about this. Let X be a path-connected, locally path-connected and semi-locally simply connected space, and let two connected covering spaces

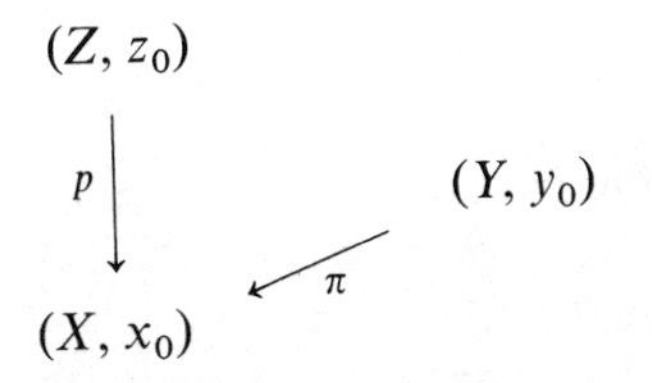

be given, whose characteristic subgroups (H for p and G for π) are contained in one another: $H \subset G \subset \pi_1(X, x_0)$. Then by the liftability criterion p can be "lifted to π", i.e. there is exactly one continuous map $f: (Z, z_0) \to (Y, y_0)$ which makes the following diagram commute:

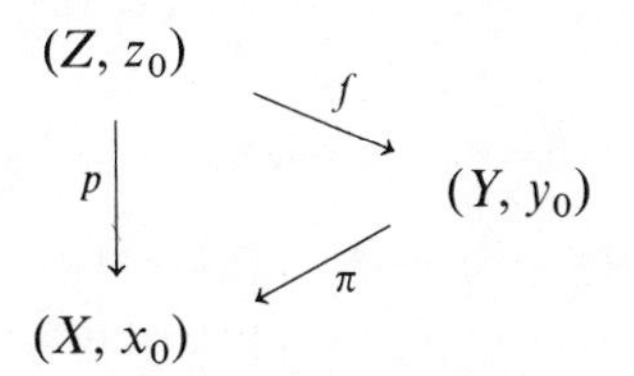

Such an f is then always a covering map; to see this, let's consider the following diagram:

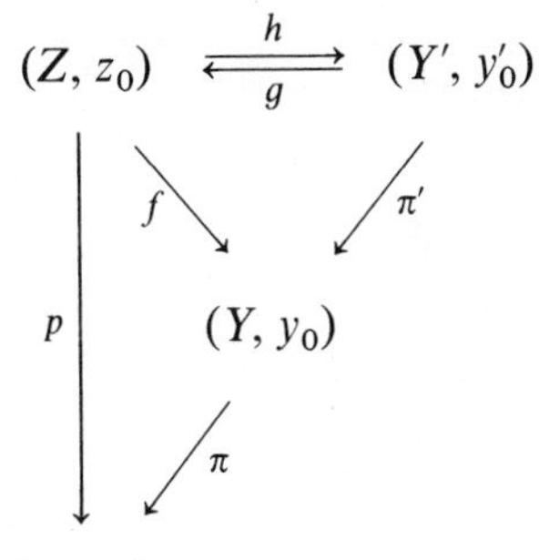

where π' is the covering map which has as characteristic subgroup the inverse image H' of H in $\pi_1(Y, y_0)$:

$$\begin{array}{ccccc} H' & \subset & \pi_1(Y, y_0) & & \\ \downarrow \cong & & \downarrow \cong & & \\ H & \subset & G & \subset & \pi_1(X, x_0) \end{array}$$

and where moreover h denotes the lifting of f to π' and g the lifting of $\pi \circ \pi'$ to p. We will show that h is an isomorphism of spaces over (Y, y_0) (and consequently that f is a covering map, since π' is). We have $h \circ f = \pi'$, and we will now verify that g is inverse to h. In any case we have $g \circ h = \mathrm{Id}_{(Z, z_0)}$, since this map is the lifting of p to itself: $p \circ g \circ h = \pi \circ \pi' \circ h = \pi \circ f = p$. To prove also that $h \circ g = \mathrm{Id}_{(Y', y_0')}$, we'll show that $h \circ g$ is a lifting of π' to itself. To do this we want to show that $\pi' \circ h \circ g$, i.e. $f \circ g = \pi'$. But this follows from the fact that both maps are liftings of $\pi \circ \pi'$ to π: The map π' is anyway, and $f \circ g$ is because $\pi \circ f \circ g = p \circ g = \pi \circ \pi'$, qed. □

To sum up: If the characteristic subgroups of two covering spaces of (X, x_0) are contained in one another, the covering space with the smaller group canonically covers the other space, and this in such a way that the three covering maps give rise to a commutative diagram

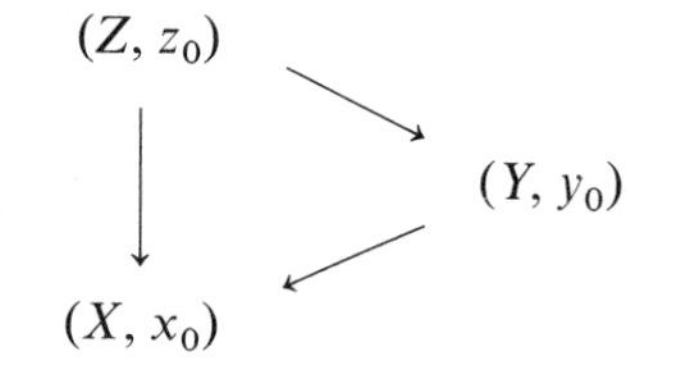

That was the preliminary consideration. Now since the characteristic subgroup of the universal cover $(\tilde{X}, \tilde{x}_0)$ is trivial, i.e. equal to $\{1\}$, it follows that:

Proposition. *The universal cover $(\tilde{X}, \tilde{x}_0)$ canonically covers every other path-connected covering space (Y, y_0) of (X, x_0), in such a way that*

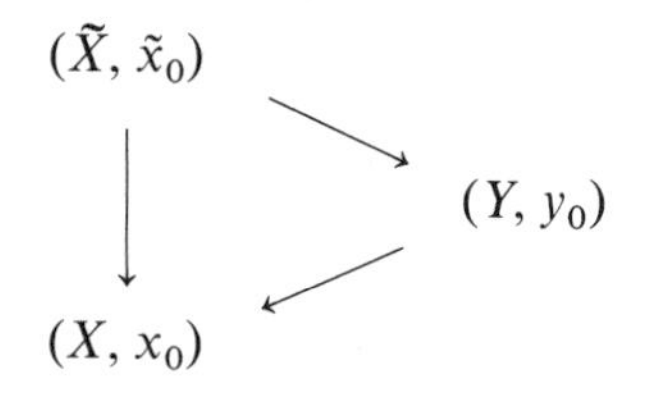

commutes.

This by itself would be sufficient reason for the universal cover to be called universal, but there is another fact that makes the point even more forcibly: The universal cover is in particular normal (obvious), and if $\mathscr{D}_X$ denotes the group of covering transformations $\tilde{X} \to X$, we obtain the canonical homeomorphism $\tilde{X}/\mathscr{D}_X \cong X$, produced by the projection itself. Choosing basepoints $\tilde{x}_0 \to x_0$, we further have a canonical isomorphism $\pi_1(X, x_0) \cong \mathscr{D}_X$, as described in detail in the theorem about the group of covering transformations. Now let's consider the two groups of covering transformations for the situation described in the above proposition:

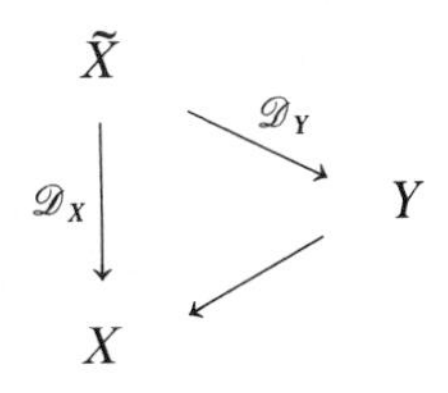

Then we have $\mathscr{D}_Y \subset \mathscr{D}_X$, and relative to $\pi_1(X, x_0) \cong \mathscr{D}_X$ the set $\mathscr{D}_Y$ is no other than the characteristic subgroup $G(Y, y_0) \subset \pi_1(X, x_0)$; and since of course $\tilde{X} \to Y$ also induces a homeomorphism $\tilde{X}/\mathscr{D}_Y \cong Y$, we obtain in all the following

Universality Theorem for the Universal Cover. *Let X be a path-connected, locally path-connected and semi-locally simply connected space, let $x_0 \in X$, and let $(\tilde{X}, \tilde{x}_0) \to (X, x_0)$ be the universal cover and $\mathscr{D}_X \cong \pi_1(X, x_0)$ the group of covering transformations of $\tilde{X} \to X$. Then if $\Gamma \subset \mathscr{D}_X$ is an arbitrary subgroup, the map*

$$\begin{array}{c}(\tilde{X}/\Gamma, [\tilde{x}_0]) \\ \downarrow \\ (X, x_0)\end{array}$$

is the covering map of a path-connected covering space, and all path-connected covering spaces of (X, x_0) are obtained in this way, up to a uniquely determined isomorphism.

*

I'd like to conclude this section with a couple of very short remarks on how the all-important fundamental groups can be calculated. One means is covering space theory itself: Now and then it is easy to determine the group of covering transformations of the universal cover of X. For instance we have $\pi_1(S^1, x_0) \cong \mathbb{Z}$, because the translations by integers $\mathbb{R} \to \mathbb{R}$ obviously form the group of covering transformations of the universal cover $\mathbb{R} \to S^1$, $x \mapsto e^{2\pi i x}$; and for $n \geq 2$ we have $\pi_1(\mathbb{RP}^n, x_0) \cong \mathbb{Z}/2\mathbb{Z}$, because the universal cover $S^n \to \mathbb{RP}^n$ has multiplicity two.

There is also the following trivial but useful observation: The fundamental group of a product is the product of the fundamental groups:

$$\pi_1(X \times F, (x_0, f_0)) \cong \pi_1(X, x_0) \times \pi_1(F, f_0)$$

in a canonical way.

Covering spaces and products are special cases of locally trivial fibrations and these are special cases of Serre fibrations, for which the "exact homotopy sequence" contains information about the fundamental group of base, fiber and total space (see, for instance, [11], p. 65); and let's not forget also that the functor π_1 is trivially homotopy invariant.

Finally I should mention the important Seifert–van Kampen theorem, which under certain conditions allows us to find the fundamental group of a space $X = A \cup B$ given the following three groups and two homomorphisms (see for instance [16], III, 5.8):

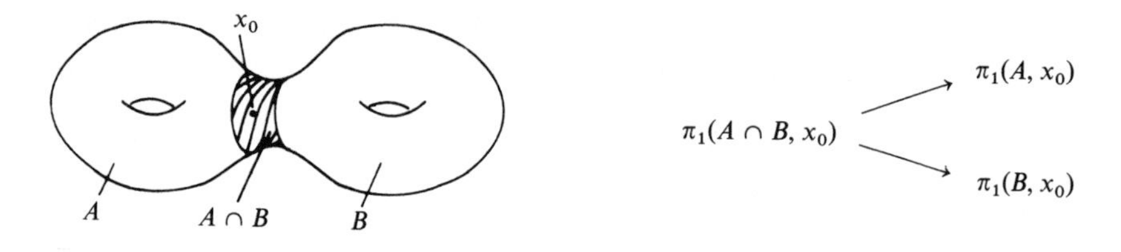

§8. The Role of Covering Spaces in Mathematics

The notion of a covering space originates from function theory, in particular from the study of "multi-valued" holomorphic functions, which arise by analytic continuation. It was discovered by Riemann at a time when there were not yet means to understand it exactly, by today's standards.

Let $G \subset \mathbb{C}$ be a domain and f a germ of a holomorphic function that can be analytically continued along any path contained in G (like for instance $\sqrt{z}$ in $\mathbb{C}\backslash 0$ or log in $\mathbb{C}\backslash 0$ or $\sqrt{(z-a)(z-b)}$ in $\mathbb{C}\backslash\{a, b\}$ and so on). Analytical continuation then defines a "multi-valued" function on G, and this is essentially a (single-valued) holomorphic function on a covering space of G,

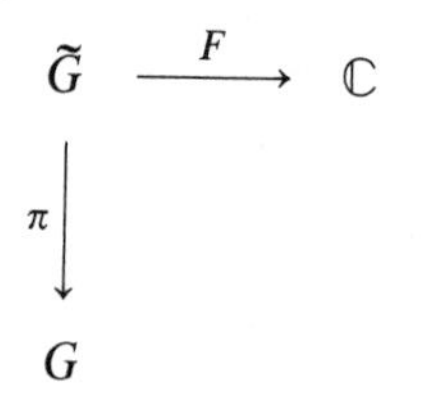

which is given "in a canonical way" (it is easy to say this in retrospect) by the continued germs.

Incidentally, the coverings thus obtained are really coverings in our sense (see definition in p. 130). Branching points only start to appear when $\tilde{G}$ is completed by adding some points of $\mathbb{C}\backslash G$ into which f cannot be analytically continued (like, for instance, 0 for $\sqrt{z}$), and covering spaces with "holes" (as the example in p. 131) come up when f can be continued to all points of G, but not along every path.

So the multi-valued functions which are encountered in function theory (and I need not dwell on the interest that functions like $\sqrt{z}$ present to the mathematician) can now be correctly understood in the covering space setting, and more: they are made accessible to the usual function-theoretical methods. This is not only the original motive for the invention of covering spaces, but also, even to this day, an important application, not having been superseded, as one might have assumed, by any more modern methods.

But things didn't stop there. Let me first make the general observation that covering spaces very often "occur in nature", that is, one comes across them spontaneously, while studying entirely different problems, and then one can thankfully pocket the seemingly heaven-sent information provided by covering space theory. Suppose for instance that a finite group G operates freely on a topological space Y; then the quotient map $Y \to Y/G$ is a covering map. Or if we're dealing with a family of differentiable functions with no bifurcation of the singularities; then the singularities form a covering of the base space. And so on and on.

But covering spaces are often intentionally introduced as tools. Covering spaces have a certain tendency of being "simpler" than the space they cover (the example $S^n \to \mathbb{RP}^n$ can be taken as a symbol of that), so the application principle is generally the following: The object of primary interest is X, but X is too complicated for a direct grasp, so one goes over to a more transparent covering space Y of X, and uses covering space theory to obtain information on X from information on Y. For instance, every non-orientable manifold M has an orientable two-fóld covering manifold $\widetilde{M} \to M$ ("orientable double cover"), and this is the vehicle to verify in the non-orientable case certain assertions, whose proof at first works "willingly" only for orientable manifolds.

In a number of applications this simplification process displays all its power only when one ascends all the way to the universal cover. I will proceed to mention three significant examples of this.

(1) Riemann Surfaces. Riemann surfaces are the connected one-dimensional complex manifolds, well-known from function theory. As topological spaces they are two-dimensional manifolds, hence surfaces. Let X be a Riemann surface and $\pi: \widetilde{X} \to X$ its universal cover. Then $\widetilde{X}$ is at first just a topological space and not a Riemann surface yet, but the complex structure of X is immediately carried over to covering spaces; it is easy to verify that there is on $\widetilde{X}$ exactly one complex structure for which π is holomorphic. So then $\widetilde{X}$ is a simply connected Riemann surface, and simply connected Riemann surfaces are in fact much easier to understand than Riemann surfaces, period: By the Riemann mapping theorem for Riemann surfaces $\widetilde{X}$ is bi-holomorphically equivalent to either the complex plane $\mathbb{C}$ or the Riemann sphere $\mathbb{CP}^1$ or the open unit disc $U \subset \mathbb{C}$! Now how does one use this knowledge to obtain information about X? Well, the covering transformations of $\widetilde{X} \to \widetilde{X}$ are bi-holomorphic maps (this is trivial and not a consequence of any special theorem); the group of covering transformations $\mathcal{D}$ operates freely and "properly discontinuously", i.e. each $\tilde{x} \in \tilde{X}$ possesses a neighborhood U such that the $\varphi(U)$, $\varphi \in \mathcal{D}$, are pairwise disjoint; the orbit space $\tilde{X}/\mathcal{D}$ of such an action has then a complex structure inherited from $\tilde{X}$, namely, the only one that makes $\tilde{X} \to \tilde{X}/\mathcal{D}$ holomorphic; and so the homeomorphism $\tilde{X}/\mathcal{D} \cong \tilde{X}$ given by covering space theory is obviously bi-holomorphic. Without using anything more sophisticated than covering space theory (a topological theory!) and the Riemann mapping theorem one can thus obtain the following: Up to bi-holomorphic equivalence the Riemann surfaces are exactly the quotients $\tilde{X}/\mathcal{D}$, where $\tilde{X} = \mathbb{CP}^1$, $\mathbb{C}$ or U, and $\mathcal{D}$ is a subgroup of the bi-holomorphic automorphisms of $\tilde{X}$ acting freely and properly discontinuously.

The groups of automorphisms of $\mathbb{CP}^1$, $\mathbb{C}$ and U have been explicitly established and well-known for a long time; the subgroups acting freely and properly discontinuously can in principle be searched for, and $\tilde{X}/\mathcal{D}$ studied—and while this is by no means a simple problem for the case $\tilde{X} = U$, one has at least a very concrete starting point for further studies, and we're one great step ahead of the original situation where all we had was "let X be a Riemann surface".

(2) Space Forms. A classical problem of differential geometry, to this day not completely solved, is the classification of *space forms*. By a space, or Clifford–Klein, form is meant a complete connected n-dimensional Riemann manifold $(M, \langle .., .. \rangle)$ with constant sectional curvature K. (See [21], p. 69.) Without loss of generality we can consider only the cases $K = +1, 0, -1$. A connected covering space of a space form is again in a canonical way a space form of the same curvature, and in analogy to the Riemann mapping theorem one has here the Killing–Hopf theorem: The sphere S^n, the Euclidean space $\mathbb{R}^n$ and the "hyperbolic space" $\mathbb{H}^n$ are the only simply connected space forms up to isometry, with curvatures $K = +1, 0$ and -1 respectively.

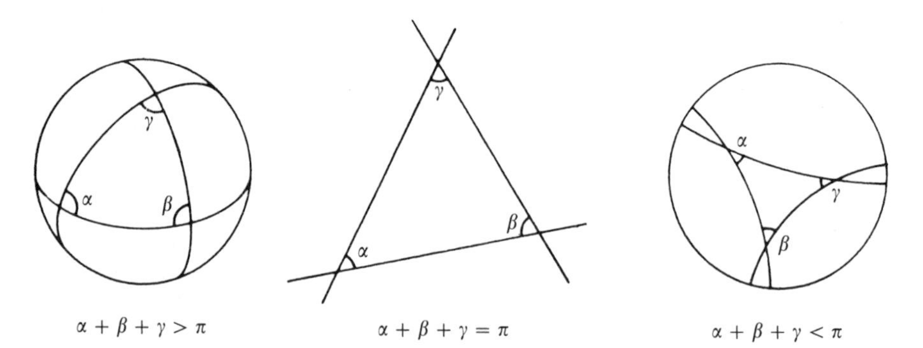

Geodesic triangles of the sphere S^2, Euclidean plane $\mathbb{R}^2$ and hyperbolic plane ($\mathring{D}^2$ with "hyperbolic metric")

The isometry groups of these three spaces are well-known, and in analogy to the case of Riemann surfaces covering space theory shows that: The quotients of S^n, $\mathbb{R}^n$ and hyperbolic space by subgroups of the isometry group acting freely and properly discontinuously are, up to isometry, the only space forms that exist.

(3) Lie Groups. A Lie group is a differentiable manifold with a "differentiable" group structure (i.e. $G \times G \to G$, $(a, b) \mapsto ab^{-1}$ is differentiable). Lie groups play an important role in various parts of mathematics and, for that matter, in theoretical physics also; $O(n)$, $GL(n, \mathbb{R})$, $GL(n, \mathbb{C})$, $SO(n)$, $U(n)$, $SU(n)$ are a few generally known examples. Covering space theory shows that the universal cover $\tilde{G}$ of a connected Lie group is again a Lie group in a canonical way, and that G is the quotient $\tilde{G}/H$ by a discrete subgroup H of the center of $\tilde{G}$. But the simply connected Lie groups are amenable to classification, since they are essentially determined by their "Lie algebra".

*

I don't want to leave you with the impression that the covering space trick is the essential point in these classification problems; just the Riemann mapping theorem by itself is already much deeper than the whole of covering space theory from A to Z. But one could say that the notion of covering space, as well as many other topological basic notions, is an indispensable concept in a number of significant contexts, and should be known by every mathematician.

CHAPTER X

The Theorem of Tychonoff

§1. An Unlikely Theorem?

Already in Chapter I, about fundamental concepts, we had convinced ourselves that the product $X \times Y$ of two compact topological spaces is again compact, and by induction it follows of course that the product of finitely many compact spaces is always compact. In VI, §2 we had also considered products of arbitrarily many factors, and we'll come back to them now, since this chapter is devoted to the following

Theorem (Tychonoff 1930). If $\{X_\lambda\}_{\lambda \in \Lambda}$ *is a family of compact topological spaces, the product* $\prod_{\lambda \in \Lambda} X_\lambda$ *is also compact.*

*

Anyone hearing the theorem of Tychonoff for the first time must admit that our intuition of the notion of compactness would suggest the opposite for infinite products. For compactness is a finiteness property (finite open covers), and so it is not surprising that it is carried over to spaces obtained by *finite* unions or products of compact spaces; but we do not expect that a construct of *infinitely* many compact elements would again have to be compact. The simplest examples show that successive enlargements of compact spaces can lead to something non-compact in the end: for instance, CW-complexes with infinitely many cells are always non-compact; non-compact manifolds can be "exhausted" by compact subsets,

or, to mention an entirely trivial, but not atypical process: Adjoining one isolated point to a compact space gives again a compact space; but doing this infinitely often, i.e. taking the disjoint union with an infinite discrete space, the result is non-compact.

From the same point of view, if one considers the sequence of "cubes"

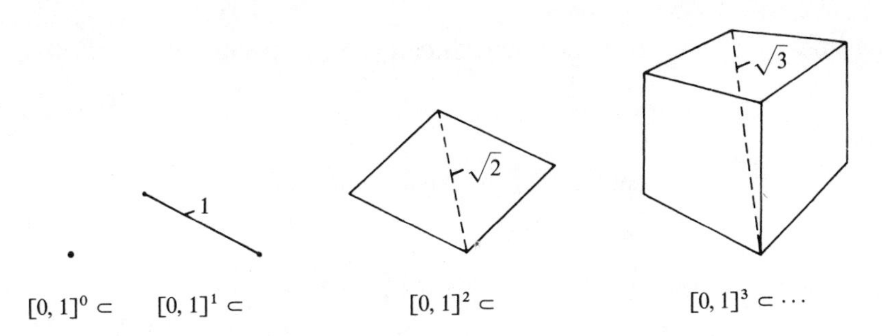

one would hardly get the feeling that $[0, 1]^\infty$ must be compact; likewise the compactness of $\{0, 1\}^\infty$ does not sound very plausible when we think of $\{0, 1\}^0 \subset \{0, 1\}^1 \subset \cdots$: Isn't $\{0, 1\}^\infty$ something very similar, if not actually the same, as an infinite discrete space?

"Against" the theorem of Tychonoff could also be adduced the fact that the unit ball in a normed space is compact only in the finite-dimensional case: one more evidence that supports the view that infinite dimension is an obstacle to compactness.

And yet here our intuition is misleading, but not so much our intuition of compactness, rather more that of products. We naturally derive our intuition of products from products in $\mathbb{R}^3$ of two or three factors, and thus it is not so conspicuous that "closeness" in the product topology of infinite products is still a condition on only finitely many coordinates: For any ever so small neighborhood U of a point $x_0 \in \prod_{\lambda \in \Lambda} X_\lambda$ the assertion $u \in U$ says absolutely nothing about most (i.e. all but finitely many) components u_λ, because U must contain a box of the form $\pi_{\lambda_1}^{-1}(U_{\lambda_1}) \cap \cdots \cap \pi_{\lambda_r}^{-1}(U_{\lambda_r})$. For this reason the image of the ∞-dimensional cube which we derive from the finite-dimensional case is not entirely appropriate. In our eyes, which always want to interpret "close" as "metrically close", the fact that components of $(x_1, x_2, \ldots)$ that lie "very far away" are relatively unimportant is much better represented by the so-called Hilbert cube: The box in separable Hilbert space whose edge lengths are $1/n$ in the direction of the e_n-axis (total diameter $\sqrt{\sum 1/n^2} = \pi/\sqrt{6}$), which can be visualized by considering the analogous lower-dimensional boxes:

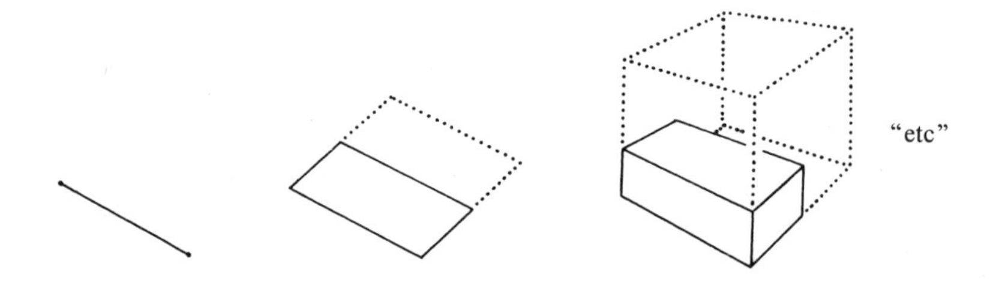

In fact the Hilbert cube is actually homeomorphic to the product of countably many intervals $[0, 1]$. (The map $(x_n)_{n \geq 1} \mapsto (x_1, x_2/2, x_3/3, \ldots)$ gives a homeomorphism from the product to the Hilbert cube, as can be easily verified.)

*

Now after having heard that the theorem of Tychonoff is true, one might think, on the grounds of previous experience with similar-sounding results, that the proof shouldn't be at all difficult: "Just the way these things always go: Let $\mathfrak{B} = \{V_\alpha\}_{\alpha \in A}$ be an open cover of $\prod_{\lambda \in \Lambda} X_\lambda$. For every $x \in V_\alpha$ the set V_α must contain a whole box $U_{\lambda_1} \times \cdots \times U_{\lambda_r} \times \prod_{\lambda \neq \lambda_i} X_\lambda$. Suppose now that there is no finite subcover. Then ... and so on." But no! ... Although many proofs in point-set topology sort of work by themselves, guided by intuition and oiled by a cunning terminology and spatial intuition—the proof of the Tychonoff theorem is not one of these.

§2. What Is It Good For?

A theorem that goes against intuition has its existence justified by this fact alone. All right. An equally general, but perhaps more weighty point of view is that every discipline must strive to clarify it own fundamental concepts.

The concepts don't just come out and advertise themselves; it is the mathematician's task to pick the most convenient among several similar concepts, and the theorem of Tychonoff, for instance, has been a decisive reason to give preference to the compactness concept defined via open covers, over the concept of sequential compactness, which is not transferred to infinite products.

But what's the situation like as regards applications outside point-set topology itself? I'd venture to assert that differential and algebraic topology make no essential use of the Tychonoff theorem. But in functional analysis the theorem comes to the fore in several very meaningful places, and I will mention three of these here. My purpose is only to show how the theorem of Tychonoff is introduced in carrying out each proof. To give these proofs in full would be pedantic, since the context in which they belong can only be delineated here.

(1) Weak Compactness of the Unit Ball in Reflexive Banach Spaces. Let X be a normed space over $\mathbb{K} = \mathbb{R}$ or $\mathbb{C}$. For a continuous linear map $f : X \to \mathbb{K}$ ("linear form") we define $\|f\| := \sup_{\|x\| \leq 1} |f(x)|$, and this makes the space X' of linear forms into a normed space, called the *dual space* of X. The dual space is always a Banach space, even when X itself is not complete.

Every element $x \in X$ canonically defines a linear form on the space of linear forms, by the formula $x : X' \to \mathbb{K}, f \mapsto f(x)$; and this actually gives an injective isometric linear map $X \subset X''$, by means of which one can always regard X as a subset of X''. X is called *reflexive* when, furthermore, $X = X''$. Hilbert spaces, for example, are reflexive.

By the *weak topology* on a normed space X we mean the coarsest topology for which the $f : X \to \mathbb{K}$ are continuous for all $f \in X'$. A subbasis is:

$$\{f^{-1}(U) \mid f \in X', U \subset \mathbb{K} \text{ open}\}.$$

On every normed space there are thus two topologies: first the topology of the norm, which is what is meant when the word topology is used by itself; and second, the weak topology. On the dual space X' one considers also a third, ever "weaker" (i.e. coarser) topology, namely the *weak-* topology*, which is the coarsest topology for which the $x : X' \to \mathbb{K}$ are continuous for every $x \in X$. A sequence $(f_n)_{n \geq 1}$ in X' is weak-* convergent if and only if it is pointwise convergent, i.e. if $(f_n(x))_{n \geq 1}$ is a convergent sequence of numbers for every x.

Corollary of the Theorem of Tychonoff. *The unit ball in X' is compact in the weak-* topology.*

Outline of Proof. Let D be the interval $[-1, 1]$ (resp. the disc $\{z \in \mathbb{C} \mid |z| \leq 1\}$) in $\mathbb{K}$, and $D_x := \{\|x\| \cdot z \mid z \in D\}$. Then we know from the Tychonoff theorem that $\prod_{x \in X} D_x$ is compact anyway, hence every closed subspace of this product is compact, and we'll see that the unit ball $U' := \{f \in X' \mid \|f\| \leq 1\}$ with the weak-* topology is homeomorphic to such a closed subspace. In fact, one defines $U' \to \prod_{x \in X} D_x$ by $f \mapsto \{f(x)\}$, $x \in X$. This map is obviously injective; the component maps $f \mapsto f(x)$ are continuous by definition of weak-* topology, so the whole map is continuous in this topology. Let $\tilde{U}$ be its image. For a fixed $x \in X$ and $U \subset \mathbb{K}$ open the set $\{f \in U' \mid f(x) \in U\}$, which is in the subbasis, is taken onto $\tilde{U} \cap \pi_x^{-1}(U)$, so the map $U' \to \tilde{U}$ is indeed a homeomorphism. Next one proves that $\tilde{U}$ is closed in $\prod_{x \in X} D_x$; this takes some work, but does not require any more advanced tool. And so the assertion follows. . . . qed. □

For reflexive spaces, of course, the weak topology on X' is the same as the weak-* topology, so the unit ball is weakly compact in X' and also in $X'' = X$, by the same argument. If X is also separable, the whole space with the weak topology is not first countable, but the unit ball is (it is even metrizable, see [4], p. 75), and so it is not only compact, but actually sequentially compact: Every bounded sequence in the norm has a weakly convergent subsequence. . . . .

(2) Compactness of the Spectrum of a Commutative Banach Algebra. A *commutative Banach algebra* is a complex Banach space B together with a multiplication law which makes it into a commutative $\mathbb{C}$-algebra with 1 and which satisfies $\|ab\| \leq \|a\| \cdot \|b\|$. The simplest and so to speak most "transparent" examples are the algebras $C(X)$ of bounded continuous functions on topological spaces X. But the more interesting examples are not so much the function algebras but *operator* algebras. The study of operators (for example, differential or integral operators) is indeed one of the main aims of functional analysis. Now if one has one or several mutually commuting operators in a Banach space, they form a commutative sub-algebra B in the (noncommutative) Banach algebra of all operators of the space, and it is quite plausible that a more accurate knowledge of B as a Banach algebra, i.e. up to an isomorphism of Banach algebras, may contain useful information about these operators. Of course the individual traits of these operators are lost when they are considered in this way: for instance, if they are differential operators, and what they act on. These are properties that cannot be inferred from the isomorphism type of the Banach algebra, just the same as when the application of an algebraic topological functor suppresses the individual traits of a geometric problem. But many important properties of the operators in the Banach algebra remain recognizable, first the algebraic ones, for instance whether the operator is a projection ($b^2 = b$) or if it is nilpotent ($b^n = 0$), or invertible, or if it has a "square root": $b = a^2$. But more than that, the norm of operators is still available in the Banach algebra, so one can consider limit processes, for instance power series of operators and so on.

But how can we materialize our wish for "insight" into the structure of the Banach algebra? Well, a high degree of insight into this structure would be obtained if one could find a topological space X and an isomorphism of Banach algebras $B \cong C(X)$! How, under what circumstances, can this be achieved? To find this out, one has to study how, and whether, a given space X can be reconstructed from the Banach algebra structure of $C(X)$. So the question is: How can a point $x \in X$ be made perceptible as a (Banach) *algebraic* object? There are actually two outward manifestations of the points of X that offer themselves. First, each x defines via $f \mapsto f(x)$ an algebra homomorphism $C(X) \to \mathbb{C}$, which characterizes x when the space X is not too unreasonable; all one need is, given two points $x \neq y$, a continuous bounded function that takes different values on the two points. So in an arbitrary commutative Banach algebra one could concentrate on the algebra homomorphisms $B \to \mathbb{C}$ as an alternative to the points $x \in X$.

On the other hand every $x \in X$ in $C(X)$ defines an ideal, the annihilator of x, given by $\mathfrak{a}_x := \{f \in C(X) \mid f(x) = 0\}$. This is obviously a maximal ideal: if an ideal contains both $\mathfrak{a}_x$ and another function f such that $f(x) \neq 0$, it must contain every function: $\mathfrak{a}_x + \mathbb{C} \cdot f = C(X)$, trivially. Again for reasonable spaces we'll have $\mathfrak{a}_x \neq \mathfrak{a}_y$ for $x \neq y$. So it would be a reasonable (though not necessarily successful) start to consider, for a commutative Banach algebra B, the so-called spectrum of B, that is

$$\text{Spec } B := \text{Set of maximal ideals of } B,$$

as a candidate for the underlying set of the desired space.

In fact both starts are two different descriptions of one and the same thing: To every algebra homomorphism $B \to \mathbb{C}$ there corresponds a maximal ideal, namely, its kernel; and this correspondence between algebra homomorphisms and maximal ideals is bijective, because, by a theorem that is not difficult to prove (Gelfand–Mazur), for every maximal ideal $\mathfrak{a}$ there is exactly one algebra homomorphism $B/\mathfrak{a} \cong \mathbb{C}$. So we can look at the elements of Spec B both as maximal ideals $\mathfrak{a}$ and as algebra homomorphisms $\varphi: B \to \mathbb{C}$, in the way indicated; and for our purposes of representing B as a function algebra, the special case $B = C(X)$ unmistakably suggests what functions we should associate to each $b \in B$, namely $f_b: \text{Spec } B \to \mathbb{C}, \varphi \mapsto \varphi(b)$.

Algebra homomorphisms $\varphi: B \to \mathbb{C}$ are automatically linear forms of norm 1, so Spec B is canonically a subset of the unit sphere in the dual space B'. In particular the functions f_b are always bounded (by $\|b\|$).

We haven't chosen a topology on Spec B yet, but if we only want all the f_b to be continuous, we'll try to achieve this in the most economical possible way, and this means exactly that we have to give Spec $B \subset B'$ the topology induced by the weak-* topology! Then we really get a canonical algebra homomorphism $\rho: B \to C(\text{Spec } B)$, $b \mapsto f_b$. Is this an isomorphism? Now, not every commutative Banach algebra is isomorphic to a $C(X)$. On $C(X)$ there is still an additional algebraic structure which must be required to be introducible in B, namely complex conjugation, as follows: By an "involution" $*: B \to B$ on a commutative Banach algebra is meant an $\mathbb{R}$-algebra homomorphism with the properties $(\lambda \cdot 1)^* = \bar{\lambda} \cdot 1$ for every $\lambda \in \mathbb{C}$ and $b^{**} = b$

and $\|b^*b\| = \|b\|^2$ for all $b \in B$. A commutative Banach algebra with involution is called a B^*-algebra. For such algebras there is the following

Theorem (Gelfand–Neumark). *If $(B, {}^*)$ is a B^*-algebra, then $\rho: B \to C(\text{Spec } B)$ is an isometric B^*-algebra isomorphism.*

So this is the answer or an answer to the question posed at the beginning. Where this question came from and where the answer leads to is a matter functional analysis has much to say about, but I think that from the little that I have indicated here you can realize I mean it when I say it: The spectrum of a commutative Banach algebra is an "important" concept of functional analysis. Now the theorem of Tychonoff makes a remarkable assertion about the spectrum. As we have already seen, Spec B is a subspace of the unit sphere of B' with the weak-* topology, which is compact by Tychonoff. It is not difficult to show that Spec B is in fact a closed subspace, and so the following result, which is especially striking in view of the Gelfand–Neumark theorem, ensues:

Corollary of the Theorem of Tychonoff. *The spectrum of a commutative Banach algebra is compact.*

*

(3) Stone–Čech Compactification. In the heuristic process of the preceding paragraph we were trying to reconstruct X from $C(X)$, but as the corollary of the theorem of Tychonoff shows, Spec $C(X)$ cannot always be equal to X, because X does not have to be compact. In what relationship to each other do X and Spec $C(X)$ stand? Without any additional assumptions the canonical mapping $X \to \text{Spec } C(X)$ doesn't have to be either injective or surjective. If it doesn't happen to be injective, however, this is the result of a rather uninteresting cause, approximately the fact that the topology on X is so coarse that the continuous bounded functions cannot separate all points. (For the trivial topology, for instance, every continuous function is constant, so Spec $C(X)$ is a point.) So in order to exclude this effect one assumes some separation property, and the right separation property here turns out to be what is called "completely regular": Points must be closed and given any closed set A and any point $p \notin A$ there must be a continuous $f : X \to [0, 1]$ with $f(p) = 0$ and $f|A = 1$; this happens for instance in every Hausdorff space in which the Urysohn lemma can be applied. But then the following theorem holds (see [8], p. 870): If X is completely regular, then the canonical mapping $X \to \text{Spec } C(X)$ is an embedding, i.e. a homeomorphism onto its image, and this image is a dense subspace, i.e. its closure is the whole of Spec $C(X)$.

By means of this embedding X itself can be considered as a dense subset of the space Spec $C(X)$ which, by Tychonoff, is compact: In particular every completely regular space is a subspace of a compact space, which in itself is already quite amazing. Spec $C(X)$ is the so-called "Stone–Čech compactification" of a completely regular space, and it is generally denoted by βX. In a certain sense it is the "biggest" compactification: it can be characterized by the property that every continuous map of X into a compact Hausdorff space can be extended to βX .... To do justice to the Stone–Čech compactification another book would be required (as well as another author), but even without that I hope to have infused you in the meantime with some respect for the Tychonoff theorem, the proof of which we're going to turn our attention to now.

§3. The Proof

All proofs of the Tychonoff theorem use "Zorn's lemma", which we are going to talk about first. After that I want to seize the opportunity to introduce the concepts of *filter* and *ultrafilter*, which are useful elsewhere as well. Equipped with these tools we will then show: If a space X has a subbasis $\mathfrak{S}$ with the property that every cover of X by sets of $\mathfrak{S}$ possesses a finite subcover, then X is already compact. So to apply this to a product $X = \prod_{\lambda\in\Lambda} X_\lambda$ of compact spaces all one has to do is prove that the canonical subbasis of cylinders $\{\pi_\lambda^{-1}(U_\lambda) \mid \lambda \in \Lambda, U_\lambda \subset X_\lambda \text{ open}\}$ has this property, and the theorem of Tychonoff will be proven. Let's start out by convincing ourselves of this property of the subbasis: Let $\mathfrak{U}$ be a cover of the product by open cylinders. Assume $\mathfrak{U}$ has no finite subcover. Then there is in every factor X_λ at least one point x_λ whose "coordinate plane" $\pi_\lambda^{-1}(x_\lambda)$ is *not* covered by finitely many sets in $\mathfrak{U}$, and this for the following reason: A coordinate plane which is covered by finitely many cylinders of $\mathfrak{U}$ always fits in *one* such cylinder, otherwise the finitely many cylinders would cover the whole product, which is against the hypothesis; but if *every* coordinate plane over X_λ fits in a cylinder of $\mathfrak{U}$ it follows from the compactness of X_λ, again contrary to the hypothesis, that a finite number of the cylinders cover the product. So for every λ there is an x_λ as asserted.

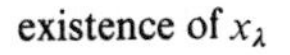

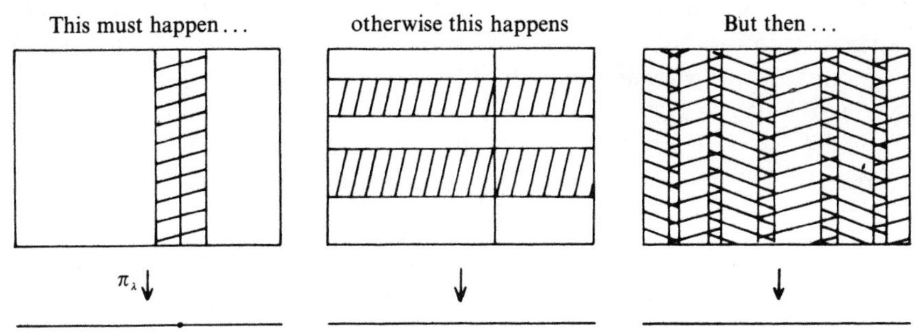

Now let $x := \{x_\lambda\}_{\lambda \in \Lambda}$. Then x must lie in some cylinder $\pi_\mu^{-1}(U_\mu)$ of $\mathfrak{U}$ and it follows that the whole coordinate plane $\pi_\mu^{-1}(x_\mu)$ is contained in the cylinder, which contradicts our construction—so the assumption was false, qed.

1. Zorn's Lemma. As you know one often has reason to consider "maximal" or "minimal" mathematical objects of a certain type. In the preceding paragraphs, for instance, we talked about maximal ideals in a commutative Banach algebra; a differentiable structure on a manifold is by definition a maximal differentiable atlas; in the theory of Lie groups the maximal compact subgroups of a connected Lie group are important; the maximal open set contained in a given subspace A of a topological space is called its interior $\mathring{A}$, and the minimal closed set containing A is its closure $\bar{A}$; the finest and coarsest topologies with certain properties are maximal and minimal in the set of such topologies; and so on and on.

In many cases, one could even say in most, the objects in question are in particular subsets of a fixed set, and the order relation which maximality or minimality refer to is inclusion of sets. Now when the property in question is transferred to arbitrary unions, the union of *all* sets with the property is naturally maximal with the property, and when the property is transferred to arbitrary intersections, the intersection of all these sets is minimal with the property. This is the very simplest situation in which the existence of maximal or minimal objects is guaranteed; the differentiable structure containing a given differentiable atlas, as well as the interior and closure of a subset of a topological space are examples of this type.

But most of the time it would be requiring to much to assume that the property is transferred to *arbitrary* unions or intersections. A substantially weaker condition, however, is often still satisfied, namely that the property be transferred to the union of intersection of *chains* of sets that possess it. This is a typical situation in which Zorn's lemma is applicable and the existence of maximal or minimal sets with the property is guaranteed.

It should be remarked at once that Zorn's lemma, too, doesn't work in all cases. For instance, to prove the existence of maximal compact subgroups in every connected Lie group, one has to go pretty deep into the theory of Lie groups; a mere formal and purely set-theoretical argument as the Zorn lemma won't do.

The Zorn lemma will be proved in the next chapter, but we will quickly preview its formulation here: As we know, a relation $\leq$ ("less than or equal to") on a set M is called a partial ordering if it is reflexive ($x \geq x$), anti-symmetric ($x \leq y$ and $y \leq x \Rightarrow x = y$) and transitive ($x \leq y \leq z \Rightarrow x \leq z$). The set $K \subset M$ is called a chain if any two elements $x, y \in K$ are related to each other, i.e. either $x \leq y$ or $y \leq x$; and K is called bounded if there is an $m \in M$ such that $x \leq m$ for all $x \in K$.

Zorn's Lemma. *If every chain in a non-empty partially ordered set* $(M, \leq)$ *is bounded,* M *has at least one maximal element, i.e. an element* a *such that there is no* $x \neq a$ *with* $a \leq x$.

2. Filters and Ultrafilters.
Definition (Filter). By a filter $\mathscr{F}$ on a topological space X (or, more generally, on any set X) is meant a set of subsets of X which satisfies the following three axioms:

Axiom 1. $F_1, F_2 \in \mathscr{F} \Rightarrow F_1 \cap F_2 \in \mathscr{F}$.
Axiom 2. $F \in \mathscr{F}$ and $F \subset F' \Rightarrow F' \in \mathscr{F}$.
Axiom 3. $\varnothing \notin \mathscr{F}$.

Definition (Convergence of Filters). A filter $\mathscr{F}$ on a topological space X converges towards a if every neighborhood of a belongs to $\mathscr{F}$.

Example. Let $(x_n)_{n \geq 1}$ be a sequence in X and let $\mathscr{F}$ be the filter of all sets in which the sequence eventually stays. Then obviously the filter converges to a if and only if the sequence does.

Definition (Ultrafilter) **and Corollary of Zorn's Lemma.** Maximal filters are called ultrafilters. Every filter is contained in an ultrafilter.

Clearly Zorn's lemma was applied here to the partially ordered set of all filters which contain the given filter.

Ultrafilters have a remarkable property:

Proposition. *If $\mathscr{F}$ is an ultrafilter on X and $A \subset X$ is a subset, then exactly one of the sets A and $X \backslash A$ belongs to $\mathscr{F}$.*

Proof. Of course both can't belong, since their intersection is empty. Moreover, one of the two has to intersect all sets in the filter, otherwise take one filter set outside A and one outside $X \backslash A$; their intersection would also be empty. Without loss of generality, let A intersect all elements of $\mathscr{F}$. Then the set of all supersets of all intersections $A \cap F$, $F \in \mathscr{F}$, is a filter containing $\mathscr{F} \cup \{A\}$, and it follows from the maximality of $\mathscr{F}$ that $A \in \mathscr{F}$, qed. □

3. Application (Proof of the Theorem of Tychonoff). So let $\mathfrak{S}$ be a subbasis of the topological space X with the property that every cover of X with sets in $\mathfrak{S}$ admits a finite subcover.

Step 1: Every ultrafilter on X converges.

Proof. Suppose there is a non-convergent ultrafilter $\mathscr{F}$. Then for every $x \in X$ we can find a neighborhood $U_x \in \mathfrak{S} \backslash \mathscr{F}$, for if all sets of $\mathfrak{S}$ containing x were elements of the filter, then so would be all finite intersections thereof, and the filter would converge towards x. Then by hypothesis $\{U_x\}_{x \in X}$ has a finite subcover: $X = U_{x_1} \cup \cdots \cup U_{x_r}$. Since the U_{x_r} are not elements of $\mathscr{F}$, their complements would have to belong to the ultrafilter, by the above remarkable property; but their intersection is empty, and we'd have a contradiction with the filter axioms, qed. □

Step 2 (the last): X is compact.

PROOF. Let $\{U_\alpha\}_{\alpha \in A}$ be an arbitrary open cover of X. Suppose there is no finite subcover, i.e. for any finite subcover there remains a non-empty "deficit" $X \setminus U_{\alpha_1} \cup \cdots \cup U_{\alpha_r}$. The set of supersets of such deficits forms then a filter; let $\mathscr{F}$ be the ultrafilter containing this filter. By step 1 we know that $\mathscr{F}$ converges towards some $a \in X$. This a must sit in some set U_α of the cover, so $U_\alpha \in \mathscr{F}$ because of convergence; but $X \setminus U_\alpha \in \mathscr{F}$, being a deficit, and this contradicts the filter axioms, qed. □

And with this we have placed the last stone in the proof of the Tychonoff theorem.

LAST CHAPTER

Set Theory

by Theodor Bröcker

This chapter is not meant to either arouse scruples or allay them. It simply summarizes, for students who have successfully concluded their first semester in mathematics, the amount of set-theoretical techniques which are occasionally useful to the mathematician.

If Λ is a set, and if every $\lambda \in \Lambda$ has a set M_λ associated to it, we define $\prod_{\lambda\in\Lambda} M_\lambda$, the *product* of the sets M_λ, as the set of maps $\varphi\colon \Lambda \to \bigcup_{\lambda\in\Lambda} M_\lambda$ such that $\varphi(\lambda) \in M_\lambda$; in other words, the product is the set of families

$$(m_\lambda \mid \lambda \in \Lambda, m_\lambda \in M_\lambda).$$

Axiom of Choice. If $M_\lambda \neq \varnothing$ for all $\lambda \in \Lambda$, then $\prod_{\lambda\in\Lambda} M_\lambda \neq \varnothing$.

So this means that if there is an element in every M_λ, there is also a function that chooses one element from each M_λ.

A *partial ordering* on a set M is a relation $\leq$ among elements of M such that the following hold: $x \leq x$ (*reflexivity*), $x \leq y$ and $y \leq x \Rightarrow x = y$ (*antisymmetry*), $x \leq y$ and $y \leq z \Rightarrow x \leq z$ (*transitivity*), in each case for all

$x, y, z \in M$. We also write $x < y$ if $x \leq y$ and $x \neq y$. If $x \in M$, $A \subset M$, we write $x \geq A$ if $x \geq a$ for all $a \in A$, and similarly $x > A$, $x < A$ etc.

Examples. If M is a set and P is the set of its subsets, inclusion defines a partial ordering on P. From this example derive the partial ordering on the subgroups of a group, on the subspaces of a vector space etc.

A *chain* or linearly ordered set is a partially ordered set in which the following holds: For every $x, y \in M$, we have either $x \leq y$ or $x \geq y$. A chain is called *well-ordered* if every non-empty subset of it has a smallest element (relative to the ordering). Example: $\mathbb{N}$, but not $\mathbb{Z}$, $\mathbb{Q}$, $\mathbb{R}$. If M, N are well-ordered, so is of course $M \times N$ with the *lexicographical* ordering, i.e. $(m, n) < (m_1, n_1)$ if $m < m_1$ or $m = m_1$, $n < n_1$. So is also $M + N$ (disjoint union) with the ordering $m < n$ for $m \in M$, $n \in N$, and as before for two elements of M or N.

In a well-ordered set the following principle holds:

Induction Principle. *If $A(k)$ is an assertion about an arbitrary $k \in K$, and $A(l)$ for every $l < k$ implies $A(k)$, then $A(k)$ for all $k \in K$.*

PROOF. Otherwise there would be a smallest $k \in K$ such that $A(k)$ is false. But then $A(l)$ for $l < k$, hence also $A(k)$. Contradiction. □

In the same way as for the natural numbers one can recursively *define* things in a well-ordered set. For instance, a recursion formula for a function f on M fixes the value $f(n)$ depending on the values $f(k)$ for $k < n$, i.e.

$$f(n) = \varphi(f \mid \{k \mid k < n\}).$$

One shows by induction on n that on the subsets $\{k \in M \mid k \leq n\}$ there is exactly one function f satisfying the recursion formula, and so the same holds on M, since a function f on M is determined by the restrictions $f \mid \{k \leq n\}$.

At this point you may hear the argument that the statement "f is uniquely determined by the recursion formula for all n" follows by induction on n. However this is not a statement of the form "For every n the following holds: ...", which can be directly proved by induction.

*

The most important tool in and from set theory is

Zorn's Lemma (essentially due to Zermelo). *Let $(M, \leq)$ be a partially ordered set. Suppose every chain $K \subset M$ is bounded. Then M has a maximal element, i.e. there is an $a \in M$ such that no $x \in M$ satisfies $x > a$.*

PROOF. Suppose the lemma doesn't hold. Then we can associate to every chain $K \subset M$ an element $m(K) \in M$ with $m(K) > K$. Here we have used the axiom of choice. Call a chain $K \subset M$ distinguished if K is well-ordered and for every *initial subset* $K_x := \{k \in K \mid k < x\}$ we have $x = m(K_x)$.

Lemma. *If K, L are distinguished chains, then either $K = L$ or $K_x = L$ or $L_x = K$ for some x in K or L respectively.*

Proof of the Lemma. Suppose we're not in one of the first two cases. Then we show by induction on K the following *assertion*: $x \in K \Rightarrow x \in L$ and $K_x = L_x$. *Proof of the assertion*: Otherwise there would be a smallest $x \in K$ for which the assertion is false. Then we already have $K_x \subset L$ (since $K_x < x$), and $K_x \neq L$ by assumption; so let $z \in L$ be minimal with the property $z \notin K_x$. Then $z > K_x$, otherwise for some $y \in K_x$ we'd have $x > y > z$; but then, since the assertion holds for y, $y \in L$ and $K_y = L_y \in z$, hence $z \in K_x$, contradicting the choice of z.

So now $z > K_x$, and clearly $K_x = L_z$. But then $x = m(K_x) = m(L_z) = z$. This proves the assertion. It now follows $K \subset L$, and since $K = L_z$ for the minimal $z \in L$ with $z \notin K$, the lemma has been shown. □

It now follows easily that the union of all distinguished chains is distinguished. Call it A. Then $m(A) > A$, and $A \cup \{m(A)\}$ is also distinguished, but then $A \cup \{m(A)\} \subset A$, a contradiction, since $m(A) \notin A$. End of the proof of Zorn's lemma. □

Definition. Two sets M, N have same *cardinality* $|M| = |N|$ if there is a bijection $M \underset{\varphi}{\rightarrow} N$. We also write $|M| \leq |N|$ if there is an injection $\varphi: M \rightarrow N$.

It is obvious that $|M| \leq |N|$ and $|N| \leq |S|$ imply $|M| \leq |S|$.

Theorem (Schröder–Bernstein).

(i) $|M| \leq |N|$ *and* $|N| \leq |M| \Rightarrow |M| = |N|$.
(ii) $|M| \leq |N|$ *or* $|N| \leq |M|$.

Proof. (i) Let $\varphi: M \rightarrow N$ and $\psi: N \rightarrow M$ be injections. We seek a bijection $\gamma: M \rightarrow N$. Every element $m \in M$ and $n \in N$ appears, up to a translation of indices, in exactly one sequence of the form

$$\cdots \underset{\psi}{\mapsto} m_{-2} \underset{\varphi}{\mapsto} n_{-2} \underset{\psi}{\mapsto} m_{-1} \underset{\varphi}{\mapsto} n_{-1} \underset{\psi}{\mapsto} m_0 \underset{\varphi}{\mapsto} n_0 \underset{\psi}{\mapsto} m_1 \underset{\varphi}{\mapsto} n_1 \mapsto \cdots$$

$n_\nu \in N, m_\nu \in M$, as an n_ν or m_ν respectively. Define $\gamma(m) = \varphi(m)$ if the sequence in which m appears starts with an $m_\nu \in M$ (and in particular has a first element), and $\gamma(m) = \psi^{-1}(m)$ otherwise. Then γ is always well-defined and bijective.

(ii) Consider the set of triples $A \underset{\varphi}{\rightarrow} B$ such that $A \subset M$, $B \subset N$, φ is bijective. Define $(A \underset{\varphi}{\rightarrow} B) < (A_1 \underset{\varphi_1}{\rightarrow} B_1)$ if $A \subset A_1$, $B \subset B_1$ and $\varphi_1 | A = \varphi$. This defines a partial ordering on the set of such triples, and every chain $((A_\lambda \underset{\varphi_\lambda}{\rightarrow} B_\lambda) | \lambda \in \Lambda)$ is bounded by $A = \bigcup_\lambda A_\lambda \underset{\varphi}{\rightarrow} \bigcup_\lambda B_\lambda = B$, $\varphi | A_\lambda = \varphi_\lambda$. Now take $A \underset{\varphi}{\rightarrow} B$ maximal by Zorn's lemma; then clearly $A = M$ or $B = N$, otherwise one could find $m \in M$, $m \notin A$, $n \in N$, $n \notin B$, and extend $A \underset{\varphi}{\rightarrow} B$ on $A \cup \{m\} \rightarrow B \cup \{n\}$ with $m \mapsto n$. □

Definition. The power set $\mathfrak{P}(M)$ is the set of subsets of M.

Theorem (Cantor). $|\mathfrak{P}(M)| > |M|$. *We also write* $|\mathfrak{P}(M)| =: 2^{|M|}$.

PROOF. Otherwise there would be a bijection $M \to \mathfrak{P}(M)$, $x \mapsto M(x)$. Define a subset $A \subset M$ by $x \in A \Leftrightarrow x \notin M(x)$. We'd have $A = M(y)$ for some $y \in M$, hence $y \in A \Leftrightarrow y \notin M(y) = A$. Contradiction. □

Theorem. *Every set can be well-ordered.*

PROOF. For a given set M consider the set of pairs (A, R), $A \subset M$, R a well-ordering on A. Set $(A_1, R_1) \leq (A_2, R_2)$ if $A_1 \subset A_2$ and $a \leq b$ in A_1 relative to R_1 if and only if $a \leq b$ relative to R_2. This defines a partial ordering among the pairs (A, R). Every chain $\{(A_\lambda, R_\lambda)\}$ is bounded by $A = \bigcup_\lambda A_\lambda, R|A_\lambda = R_\lambda$. A maximal element (A, R) satisfies $A = M$, for otherwise $m \in M$, $m \notin A$, and $A \cup \{m\}$ would be well-ordered by the well-ordering on A and the condition $A < m$, so the pair $(A \cup \{m\}, \leq)$ would be greater than (A, R). □

In the same way that cardinals are formed from sets and injections, *ordinals* are obtained from well-ordered sets and monotonic injections. Two well-ordered sets have same ordinal if there is an order-preserving bijection between them.

Theorem. *Let M, N be well-ordered. Then there is exactly one monotonic bijection from one set onto the other or an initial subset of the other. In particular the ordinals are linearly ordered.*

PROOF. Suppose there is no monotonic bijection $M \to N$ or $M \to N_x$. Then define $\varphi: N \to M_y$ inductively: if φ is already defined on N_x and $\varphi(N_x) = M_z$ for some $z \in M$, then set $\varphi(x) = z$; if $N_x \cup \{x\} \neq N$, then $N_x \cup \{x\}$ is also an initial segment in N and $\varphi(N_x) \cup \{z\}$ an initial segment in M. $\varphi(N)$ is obviously inductively defined, and $\varphi(N) = M_y$ with y minimal in M, so $y \notin \varphi(N)$. qed. □

If in particular M is well-ordered, then the cardinals $< |M|$ are represented by subsets of M, hence also by initial segments M_x of M (by the theorem), and $|M_x| \leq |M_y| \Leftrightarrow x \leq y$. So:

Corollary. *There is exactly one bijective monotonic map from the set of cardinals $< |M|$ onto an initial segment of the well-ordered set M. In particular the set of cardinals $\leq |M|$ is well-ordered by their ordering, and $|M|$ is represented by the set of ordinals $\leq$ ordinal of M.*

Theorem. *For an infinite set M we have $|M \times M| = |M|$ and $|M + M| = |M|$, where $+$ is disjoint union.*

Corollary. *If $|M|$ is infinite, $N \neq \varnothing$, then*

$$|M \times N| = |M + N| = \max\{|M|, |N|\}.$$

PROOF. From the first assertion it follows that

$$|M| = |M \times M| \geq |M \times \{1, 2\}| = |M + M| \geq |M|;$$

so the second assertion holds for the same cardinal. *Proof of the first assertion*: Consider the set of pairs (B, ψ) where $B \subset M$ is infinite and $\psi: B \to B \times B$ is bijective. If $|B| = |\mathbb{N}|$, there certainly is a bijection as required (enumeration of $\mathbb{N} \times \mathbb{N}$). As usual we're considering among the pairs (B, ψ) the ordering $(B, \psi) \leq (B_1, \psi_1)$ if $B \subset B_1$ and $\psi = \psi_1 | B$. Now Zorn's lemma provides a maximal pair (A, φ), $\varphi: A \to A \times A$. Suppose $|A| < |M|$; then $M = A + B$ and $|B| > |A|$ by the induction hypothesis (induction on the cardinality). So $M = A + A_1 + C$, $|A_1| = |A|$. Now

$$(A + A_1) \times (A + A_1) = (A \times A) + (A \times A_1) + (A_1 \times A) + (A_1 \times A_1),$$

and by the induction hypothesis there is a bijection

$$A_1 \underset{\varphi_1}{\rightarrow} (A \times A_1) + (A_1 \times A) + (A_1 \times A_1).$$

So φ_1 gives an extension of φ, namely a bijection

$$A + A_1 \to (A + A_1) \times (A + A_1),$$

which coincides with φ on A, contradicting the maximality of φ. This proves the theorem. □

Let $|M|$ be infinite and K be the set of cardinals κ such that

$$|M| < \kappa < 2^{|M|}.$$

From the first corollary of the previous theorem we obtain the following estimate:

$$0 \leq |K| \leq 2^{|M|}.$$

The *continuum hypothesis* of Cantor says that $|K| = 0$. By a theorem of Cohen this hypothesis is independent of the axioms of set theory, and inside the estimate above all assumptions are consistent with the axioms of set theory. The continuum hypothesis would imply there is no cardinal between $|\mathbb{N}|$ and $|\mathbb{R}|$; hence its name.

References

[1] Bourbaki, N., *Eléments de Mathématique*, Livre V: Espaces Vectoriels Topologiques, Chaps. I and II, 2nd ed., Paris: Hermann, 1966.

[2] Bourbaki, N., *General Topology*, Vols. I and II, Paris: Hermann and Reading, Mass.: Addison-Wesley, 1966.

[3] Bröcker, Th. and Jänich, K., *Introduction to Differential Topology*, Cambridge: Cambridge University Press, 1982.

[4] Dieudonné, J. A., *Treatise on Analysis*, Vol. II, New York and London: Academic Press, 1970.

[5] Dold, A., *Lectures on Algebraic Topology*, Berlin–Heidelberg–New York: Springer-Verlag, 1972.

[6] Dold, A., Partitions of unity in the theory of fibrations, *Ann. of Math.*, **78**, (1963), 223–255.

[7] Dunford, N. and Schwartz, J. T., *Linear Operators*, Part I: General Theory, New York: Interscience, 1957.

[8] Dunford, N. and Schwartz, J. T., *Linear Operators*, Part II: Spectral Theory, New York: Interscience, 1963.

[9] Forster, O., *Lectures on Riemann Surfaces*, New York–Heidelberg–Berlin: Springer-Verlag, 1981.

[10] Grauert, H. and Remmert, R., *Theory of Stein Spaces*, New York–Heidelberg–Berlin: Springer-Verlag, 1979.

[11] Hilton, P. J., *An Introduction to Homotopy Theory*, Cambridge: Cambridge Univ. Press, 1953.

[12] Jänich, K., *Einführung in die Funktionentheorie*, Berlin–Heidelberg–New York: Springer-Verlag, 1977.

[13] Köthe, G., *Topological Vector Spaces*, I, Berlin–Heidelberg–New York: Springer-Verlag, 1966.

[14] Milnor, J., *Morse Theory*, Princeton, NJ: Princeton Univ. Press, 1963.

[15] Neumann, J. v., Zur Algebra der Funktionaloperationen und Theorie der Normalen Operatoren, *Math. Annalen*, **102** (1930), 370–427.

[16] Schubert, H., *Topology*, Boston: Allyn and Bacon, 1968.

[17] Steen, L. A. and Seebach, J. A., *Counterexamples in Topology*, 2nd ed., New York–Heidelberg–Berlin: Springer-Verlag, 1978.

[18] Thom, R., Quelques propriétés globales des variétés différentiables, *Comm. Math. Helv.*, **28** (1954), 17–86.

[19] Tychonoff, A., Über die topologische Erweiterung von Räumen, *Math. Annalen*, **102** (1930), 544–561.

[20] Tychonoff, A., Ein Fixpunktsatz, *Math. Annalen*, **111** (1935), 767–780.

[21] Wolf, J. A., *Spaces of Constant Curvature*, New York: McGraw-Hill, 1967.

Table of Symbols

$[a, b]$	closed interval from a to b	3
$\mathring{B}$	interior of set B	6
$\bar{B}$	closure of B	6
$K_\varepsilon(x)$	ball of radius ε around a point x of a metric space	8
$+$	disjoint union, topological sum	10
$\cong$	"isomorphic"; used here for homeomorphisms	13
$(2, 3)$	open interval from 2 to 3 (I haven't got accustomed to the disgusting notation $]2, 3[$ yet; I will some day, people get accustomed to everything.) Danger of confusion with the ordered pair $(2, 3) \in \mathbb{R}^2$	14
$\lVert..\rVert$	norm	26
$\lvert..\rvert$	seminorm	27
$C(X)$	Banach space of bounded continuous functions on X	29
$[x]$	equivalence class	31
$X/\sim$	set or space of equivalence classes relative to the equivalence relation $\sim$ on X	31
G/H	quotient space of G by the subgroup H	34
X/G	orbit space of a G-space X	38

X/A	quotient space obtained by collapsing $A \subset X$ to a point	40
CX	cone over X	40
$X \vee Y$	wedge product $X \times y_0 \cup x_0 \times Y \subset X \times Y$	42
$X \wedge Y$	smash product $X \times Y/X \vee Y$	42
$Y \cup_\varphi X$	quotient space obtained from $X + Y$ by identifying x with $\varphi(x)$ ("attaching X to Y via φ").	43
$M_1 \# M_2$	connected sum	45
$X \times [0, 1]/\alpha$	quotient space obtained from $X \times [0, 1]$ by identifying $(x, 0)$ with $(\alpha(x), 1)$	46
$(\hat{X}, \hat{d})$	completion of the metric space (X, d)	51
$C_0^\infty(\mathbb{R}^n)$	vector space of C^∞-functions with compact support	56
$\simeq$	homotopic	59
$\simeq$	homotopy equivalent	61
$[X, Y]$	set of homotopy classes of maps from X into Y	61
$\pi_n(X, x_0)$	n-th homotopy group of (X, x_0)	75
$\prod_{\lambda \in \Lambda} X_\lambda$	product of family $\{X_\lambda\}_{\lambda \in \Lambda}$ of sets or topological spaces	80
$s(v_0, \ldots, v_k)$	simplex, convex hull of points $v_0, \ldots, v_k$ in general position in $\mathbb{R}^n$	88
$\|K\|$	set underlying a simplicial complex	90
X^n	n-skeleton of a cell decomposition of X	96
TM	tangent bundle	117
Y_x	fiber of a topological space $Y \xrightarrow{\pi} X$ over X at point x	128
$Y\|U$	restriction of a topological space over X to $U \subset X$	128
$\cong$	isomorphic; used here for homeomorphisms "over X"	129
$\sim$	written above a map: used in Chapter IX mostly for "liftings" of various sorts	133
α^-	path described backwards, i.e. $\alpha^-(t) := \alpha(1 - t)$	139
$\Omega(X, x_0)$	set of loops in X at x_0	140
$\simeq$	here: homotopy of loops with fixed endpoints x_0	141
$[\ ]$	here: equivalence classes of loops by $\simeq$	141
$\pi_1(X, x_0)$	fundamental group	141
f_*	homomorphism between fundamental groups induced by f	141
$G(Y, y_0)$	characteristic subgroup of $\pi_1(X, x_0)$ for the covering space $(Y, y_0) \to (X, x_0)$	142
$[\ ]_\sim$	(pages 145–149 only) equivalence classes of relation defined on page 145	145
$V(U, y)$	(pages 146–149 only—special notation required inside a proof)	146
$\mathcal{D}$	group of covering transformations	149
N_B	normalizer of subgroup B	150
$(\tilde{X}, \tilde{x}_0)$	universal cover of (X, x_0)	152
X'	dual space of a normed space X	163
Spec B	spectrum of a commutative Banach algebra	165

βX	Stone–Čech compactification of X 167
$\mathscr{F}$	filter 169
$\|M\|$	cardinality of set M 173
$2^{\|M\|}$	cardinality of power set of M 174
$\|\mathbb{N}\|$	cardinality of set $\mathbb{N}$ of natural numbers; also denoted by $\aleph_0$ (aleph null) 175

Index

The short explanations accompanying the entries are not always complete and do not replace the exact definitions given in the text.

Undergraduate Texts in Mathematics

continued from ii

Owan: A First Course in the Mathematical Foundations of Thermodynamics
1984. xvii, 178 pages. 52 illus.

Prenowitz/Jantosciak: Join Geometrics: A Theory of Convex Set and Linear Geometry.
1979. xxii, 534 pages. 404 illus.

Priestly: Calculus: An Historical Approach.
1979, xvii, 448 pages. 335 illus.

Protter/Morrey: A First Course in Real Analysis.
1977. xii, 507 pages. 135 illus.

Ross: Elementary Analysis: The Theory of Calculus.
1980. viii, 264 pages. 34 illus.

Sigler: Algebra.
1976. xii, 419 pages. 27 illus.

Simmonds: A Brief on Tensor Analysis.
1982. xi, 92 pages. 28 illus.

Singer/Thorpe: Lecture Notes on Elementary Topology and Geometry.
1976. viii, 232 pages. 109 illus.

Smith: Linear Algebra.
1978. vii, 280 pages. 21 illus.

Smith: Primer of Modern Analysis
1983. xiii, 442 pages. 45 illus.

Thorpe: Elementary Topics in Differential Geometry.
1979. xvii, 253 pages. 126 illus.

Troutman: Variational Calculus with Elementary Convexity.
1983. xiv, 364 pages. 73 illus.

Whyburn/Duda: Dynamic Topology.
1979. xiv, 338 pages. 20 illus.

Wilson: Much Ado About Calculus: A Modern Treatment with Applications Prepared for Use with the Computer.
1979. xvii, 788 pages. 145 illus.